The Great Oil Age

The Petroleum Industry in Canada

Peter McKenzie-Brown, Gordon Jaremko,
David Finch

Detselig Enterprises Ltd.
Calgary, Alberta

THE GREAT OIL AGE

© 1993 Petroleum History Society

Canadian Cataloguing in Publication Data

McKenzie-Brown, Peter
The great oil age

Includes bibliographical references.
ISBN 1-55059-072-3

1. Petroleum industry and trade—Canada—History. I. Jaremko, Gordon.　II. Finch, David, 19**-　III. Title.
HD9574.C22M32　1993　　338.2'728'0971　　C93-091745-6

Detselig Enterprises Ltd.
210, 1220 Kensington Rd. NW
Calgary, Alberta T2N 3P5

All rights reserved. No part of this book may be reproduced in any form or by any means without permission in writing from the publisher.

Printed in Canada　　ISBN 1-55059-072-3　　SAN 115-0324

£22.95

Contents

Geological Time Scale	*v*
Acknowledgments	*vii*
About the Authors	*viii*
Foreword	*ix*
Chapter 1 – The Great Oil Age	11
Chapter 2 – The Source	19
Chapter 3 – The Early Years	27
Chapter 4 – Oil Between the Wars	35
Chapter 5 – The Leduc Era	45
Chapter 6 – The Pipeline Era	49
Chapter 7 – Processing Gas	61
Chapter 8 – Oil Sands and the Heavy Oil Belt	69
Chapter 9 – Frontiers of Muskeg, Ice and Water	87
Chapter 10 – From Crude to Refined	101
Chapter 11 – Petrochemicals (The Miracle Workers)	115
Chapter 12 – A Matter of Policy	127
Chapter 13 – The Industry and the Environment	143
Chapter 14 – The Impact of the Petroleum Industry on Canada	155
Endnotes	165
Bibliography	173
Timeline	179
Index	185

Geological Time Scale

Million Years	Period		Era	Eon
− .007	Recent	Quaternary	Cenozoic	Phanerozoic
− 2.5	Pleistocene			
− 65	Tertiary			
− 135	Cretaceous		Mesozoic	
− 195	Jurassic			
− 225	Triassic			
− 280	Permian		Paleozoic	
− 320	Pennsylvanian	Carboniferous		
− 345	Mississippian			
− 395	Devonian			
− 440	Silurian			
− 500	Ordovician			
570	Cambrian			

Time Scale Distorted for Legibility

Credits:

Cover photo by David Campion

Title page photo: Oil City Royalties No. 1 well, Waterton Lakes area, Alberta, 1932. Smaller rig was from an attempt in 1919 to 1922. (Glenbow Archives/NA-700-3)

Authors photo by Andrew Jaremko

Back cover map reproduced with permission from the Geological Survey of Canada

Financial support for Detselig Enterprises Ltd. 1993 publishing program is provided by the Department of Communications, Canada Council and The Alberta Foundation for the Arts, a beneficiary of the Lottery Fund of the Government of Alberta.

Acknowledgments

This book was nearly ten years in the making and many, many people lent a hand. The book began with the Canadian Petroleum Association's public affairs group, who wanted to write a definitive history of Canada's petroleum industry. The CPA hired Fred Stenson to begin preparation of the document and he gave shape to the present product. He developed the book's structure and wrote early drafts of most chapters.

In succeeding years many others contributed to the manuscript. Notable among these were CPA vice-president Hans Maciej, who critiqued those early chapters paragraph by paragraph with one of the authors. The CPA's Christine Wignall helped research early drafts and CPA statistician Stephen Rodrigues provided numbers, dates and other information. In the end, however, the CPA took the project off its list of priorities, turning the manuscript over to the Petroleum History Society.

The present authors then took over the project, essentially rewriting the book as they brought it up to date. During the rewriting, many people commented on one or more chapters. From *Oilweek* magazine, these include Alex Rankin and the late Les Rowland; from PanCanadian Petroleum, Micky Gulless; from Shell Canada, Clint Tippett. Others who contributed their time and specialties included Bob Hutchins, Terry Robinson, Ivan Donald and A.T. Duguid. To Sherry Wilson McEwen, our editor at Detselig Enterprises Ltd., we offer our appreciation and thanks.

We have undoubtedly left out the names of others who offered comment as this book came to life and we regret those omissions. But for the help and generous contributions of our many supporters as we struggled with this book, we extend our warmest thanks.

About the Authors

Peter McKenzie-Brown has a baccalaureate in philosophy and has done graduate work in political science. Since 1977, he has worked in the public affairs departments of two petroleum-related organizations in Calgary, Gulf Canada and the Canadian Petroleum Association. He was a writer and the editor of three petroleum industry magazines, including the Canadian Petroleum Association *Review, Arctic Petroleum Review* and the *OSCAR Report*. He now edits the *Forestry, Oil & Gas Review*, a Calgary-based magazine.

Gordon Jaremko holds an M.A. in history from the University of Calgary. He has written extensively about the petroleum industry for more than ten years for the Southam news organization and is currently a writer and the resources editor for the *Calgary Herald*. He is also a regular contributor to two magazines, *Natural Gas Intelligence* in Washington D.C. and *Natural Gas Exporter* in Calgary.

David Finch is an independent, consulting historian with a M.A. in Canadian history from the University of Calgary. In 1985 he wrote *Traces Through Time,* a history of geophysics in the petroleum industry, and in 1989 a book titled *Dealmakers,* a review of the role of landmen in the petroleum industry and the history of their association. He also writes about other aspects of the history of the Canadian west.

Gordon Jaremko, David Finch and Peter McKenzie-Brown

Foreword

In the Spring of 1992, David Finch asked me if The Petroleum History Society would be interested in publishing a definitive history of the petroleum industry in Canada. My answer was a resounding "Yes!"

For several years, staff of the Canadian Petroleum Association, primarily Peter McKenzie-Brown and Hans Maciej, had been working on the manuscript for a book tentatively titled: *The Great Oil Age – The Petroleum Industry in Canada*. It was still a long way from being ready for publication and the project was losing momentum. On the condition that none of the authors would receive direct monetary gain from its publication, the Canadian Petroleum Association assigned the copyright of the manuscript and all associated rights and revenues from its publication to The Petroleum History Society. Revenues from the sale of the book are to be used solely for the advancement of the Society's primary objective – to promote activities that increase awareness of the history of the petroleum industry in Canada.

Timing of the publication of this book could not have been better. The petroleum industry in Canada stands on the threshold of a resurgency after almost a decade of severe downsizing, reorganization, and retrenchment. It would appear that the era of the large multinational oil companies controlling the Canadian petroleum industry has come to an end. The surviving large- to mid-size companies are leaner and more focused on improving the returns of already developed production. A host of small, independent, quick-acting junior oil companies has sprung into existence, aggressively exploring for new pools and plays.

Many of the new firms in Canada's petroleum industry are carrying Canadian technology abroad, particularly into countries of the former Soviet Union and the Middle and Far East. This technology, developed specifically to meet the challenges of the rugged Canadian climate, has proven adaptable to other situations and is second to none in the world.

As the petroleum industry embarks on this new era, it is appropriate to pause and reflect on all that has transpired since the first oil well at Oil Springs, Ontario, in 1858. These events and the people who shaped them have made the petroleum industry in Canada the social and economic force it is today. *The Great Oil Age – The Petroleum Industry in Canada* provides a comprehensive, authoritative and accurate account of this history and its contributions to all aspects of Canadian life.

The Petroleum History Society sincerely hopes that you enjoy reading this book as much as we have enjoyed bringing it to you. We offer our sincere appreciation to the authors, David Finch, Peter McKenzie-Brown and Gordon Jaremko, for their efforts. And we gratefully acknowledge the immeasurable contributions every man and woman who has ever worked in this industry has made to its incomparable story.

William R. S. McLellan
President, The Petroleum History Society

Chapter 1

The Great Oil Age

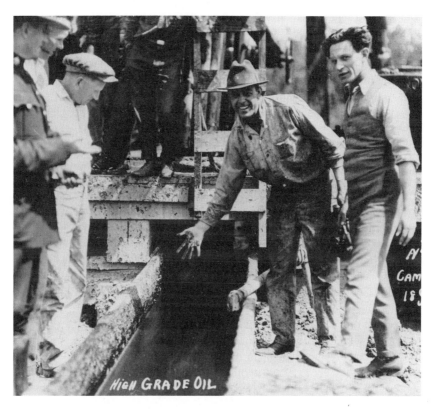

Discovery of oil, British Petroleum No. 3B at Wainwright, Alberta, July, 1925 (Glenbow Archives/NA-544-116)

The story of Canada's petroleum industry is a story of courage and peril, success and failure, wisdom and folly, high risks and high stakes, innovation, boneheadedness and frustration. The world's first oil field, located in southern Ontario, helped create an industry. In time that industry became the world's biggest and richest. But since its tenuous beginnings, it has undergone many boom and bust cycles on a trend line that once seemed to point resolutely upward. Influential since the 1850s, Canada's industry is now a leading world repository of advanced technology and expertise.

The world watched spellbound in the winter of 1991 as the might of the earth's industrial countries put a quick end to the territorial ambitions of Iraq. Toward the end of the conflict, Iraq had turned to sabotage by dynamiting

Kuwait's oil fields, setting 736 wells on fire and turning the desert into a sooty inferno. In the aftermath of the Gulf War, Kuwait called in an international array of firefighters, which allowed Canada to demonstrate the power of Canadian technology in time of peace.

Four large American firms had been on the job for two months when a Canadian well control company arrived on the scene. Calgary-based Safety Boss extinguished 180 of the oil well fires – the largest total by far. Of 14 other teams struggling to control the wells, the second best, an American outfit, snuffed out 126 fires.[1] This episode illustrates Canada's impact on the world's oil industry. Canadian influence has been greater in some ways than that of the entire Middle East.

Canada's oil and gas industry is a giant with room to grow, having already left a real mark in knowledge and technology for the last 150 years. Canadians helped define the profound difference oil and its sister commodity, natural gas, have made to human society.

THE BIRTH OF AN INDUSTRY

In the nineteenth century, oil's primary use had been for lamp fuel, candles and grease for home and factory. But the electric light bulb soon replaced the smoky lamp. Illuminating oils, which had been far more important petroleum products than gasoline, soon became of marginal importance. The change came during the battles of the First World War.

"In the oil barrel . . . is peace and war,"[2] claimed Canadian humorist Stephen Leacock in a 1930 essay, "In praise of Petroleum." Leacock was not the first to recognize that war and oil mix. For example, in a popular study of the Canadian petroleum industry which hit the bookstores during the First World War, author Victor Ross wrote, "In the greatest of wars . . . we read of armies striking at or tenaciously defending territories, for the main reason that petroleum abounds therein."[3]

By 1930 oil fueled cars, trucks, tractors, airplanes and some industrial equipment in Canada and the other industrial countries. Celebrating what he called "the Great Oil Age, which now is," Stephen Leacock wrote about "the spluttering of the farm motor on the country road, the grinding of the irresistible tractor, and long processions on the highway with motorfuls of baskets, tents, mattresses and children overflowing at the windows." The process by which oil and gas would transform society had clearly begun. Soon, these commodities would become more powerful forces in Canada's social life than Leacock or anyone else could have imagined. Oil would shape cities and towns and the lives of people.

Yet by today's standards, all the motors in all those vehicles and tools had created only a tiny market for fuels and lubricants. In the country, horse-drawn wagons were still common, and steam-powered railways were by far the most important long-distance transportation system. Although the shift from coal to fuel oil was well advanced, the diesel locomotive would

THE GREAT OIL AGE

not become the railroad's workhorse until the 1940s. As on the rails, so on Canada's lakes, rivers and seas – steamers and ocean liners would not be completely oil-fuelled until much later.

Despite a fourfold increase in automobiles within a decade, there were only a million cars on Canada's small network of mostly unpaved roads in 1930. Many farmers, unable to afford gasoline because of the worsening Depression, would hitch their autos to teams of horses. The resulting Bennett Buggy became a vivid and enduring symbol of the Dirty Thirties.

Clean-burning gas and oil furnaces were largely conveniences of the future. Chimneys and stacks belched black smoke from burning coal, which remained overwhelmingly Canada's most important source of energy until well after the Second World War.

In the thirties, the essential fuel problem was finding ways to get coal from mines in the Maritimes and the west to central Canadian furnaces and boilers.[4] One study of the day's energy problems devoted only 6 out of 140 pages to crude oil and natural gas.[5] The reason, of course, is that in those days the entire nation used only 30 million barrels of oil per year. (That volume is much less than tiny Ireland consumes today.)

In 1930, Canada imported all but three percent of the nation's oil from the United States,[6] and no national or local policies encouraged petroleum development. Debates raged about the reliability of American oil supplies. As early as 1918, thoughtful analysts had forecast the depletion of that country's oil fields within 20 years.[7] A decade later, vast discoveries had put huge surpluses of oil onto the world market. In the mid-thirties, Alberta tried to induce central Canadian businessmen to invest in the risky oil business, but found no takers.[8]

At this time, Canada's oil and gas industry was a regional business largely centred in Alberta, existing on the fringes of the national economy. After an Ottawa-financed subsidy[9] of oil production ended in 1925, the industry had no protective tariffs or other development incentives. Only after three years of Depression, in 1933, did the federal government impose a stiff tariff on refined products to encourage domestic refining. That was of no benefit to oil or natural gas producers, though – refineries built behind the tariff wall also relied mostly on foreign oil.

Seventy-two years had elapsed since Upper Canada gave birth to the world's petroleum industry. With the opening of the oil barrel, Leacock declared, "the contents flew out all over the world, like the contents of Pandora's box." But the opening of the barrel was incomplete: nylon was still the miracle fibre of the future; even the word "petrochemical" would not emerge until chemists began developing synthetic rubber for World War II. Canada's farmers began using petroleum-based fertilizers during that war. But the widespread use of plastics and the medical applications of petroleum were still far in the future.

After the Second World War came the discovery of large new oil and gas fields in Alberta, Saskatchewan and northeastern British Columbia.

Smaller fields in Manitoba made that province a modest producer. With those finds, Canada's petroleum industry became a mighty economic force. By 1990 that industry produced 550 times as much oil as in 1930, 165 times as much natural gas.[10] And Canada's producing provinces, especially Alberta, soon found their economies driven by powerful petroleum engines.

From a poor, political backwater in the Depression years, Alberta watched its national presence grow, its wealth multiply and its population triple. Petroleum royalties flowed into the provincial treasury, the economy diversified and Albertans came to enjoy the nation's highest standard of living.

PETROLEUM TECHNOLOGY

The industry's real story parallels national growth, from a small nation reliant on imported oil to an industrial nation with few peers. That story includes the metamorphosis of society as oil and gas became the dominant sources of energy. It is about the enormous industrial effort which built plants and pipelines and refineries to feed a growing hunger for hydrocarbons and their products. It is about the birth and exuberant adolescence of the modern industry in the west. It is about industrial wizardry yielding one of the world's finest oils from the tarry gunk in Alberta's oil sands. It is about the march of drilling rigs to the Arctic and the plunge into the submerged shelves of the continent – west, east, and north.

Canada is a large producer, with expertise outdistancing the magnitude of even production. Canada's technological impact far exceeds that of Russia, the world's largest oil producer and second largest natural gas producer. Canada also has more sophisticated expertise than the countries of OPEC combined. Second in strength to the enormous expertise of the United States and perhaps third to the technical power of Britain, Canada has brilliantly served world petroleum exploration and development.

Partly, that is because Canada has developed the world's most diverse petroleum industry. In doing so, this country has been host to many industry firsts. Canada has also been home to people who have become industrial legends.

For example, in the 1840s Nova Scotian Abraham Gesner developed a refining process that transmuted coal, natural tar and eventually oil into kerosene. Of equal importance, he helped found the world's first refinery – an operation that extracted kerosene from coal. In the move from whale oil lamps to petroleum lighting, this was a critical development. Until the First World War, most refined oil simply burned in lamps. Gesner's work lit homes and factories throughout the world.

Slightly after Gesner's early efforts, another Canadian dug the world's first commercial oil well. In 1858, James Miller Williams brought a well into production in Oil Springs, Ontario – a well that signalled the birth of the petroleum era. Yet most historians credit the world's first well to Edwin

Drake, an American who drilled his famous wildcat in Pennsylvania the following year.

Only three years after the Williams discovery, Thomas Sterry Hunt of the Geological Survey of Canada came up with a radical theory to explain how oil and gas fields form. Not until years later, when an American geologist came up with the same theory, did the idea gain wide acceptance.[11]

For more than 100 years, Canadians have drilled around the world. One of Canada's early drillers, William McGarvey, founded a business empire on large oil fields in Galicia (since divided between Poland and Ukraine.) His interests included the largest refinery in Europe when, during the bitter struggles of the First World War, the guns of Austria and Russia laid waste his lifetime's work.

Canadians think of oil and gas as a western industry, yet the search for these resources has affected every province and the territories. Ontario is the petroleum industry's cradle, and there was a producing oil well within the metropolitan limits of Toronto at least as recently as 1978.[12]

The Northwest Territories hosted the Norman Wells oil discovery in 1920 – Canada's largest until 1947 and still an important source of oil. And the Canol pipeline – one of the remarkable industrial projects of the Second World War – took Norman Wells oil to a gleaming new refinery in the Yukon. Located in Whitehorse, that refinery briefly piped fuel for the Pacific fleet to Skagway, Alaska.

Quebec and New Brunswick have also played roles. Oddly enough, one of Canada's most spectacular natural gas blowouts occurred on the grounds of a monastery in Quebec. In New Brunswick the colony, entrepreneurs drilled an oil and gas well in 1859; by 1909 the province produced both oil and gas. Still a producing region, New Brunswick also hosts Canada's largest refinery.

The other Atlantic provinces have been sites for offshore drilling. During the Second World War, the industry drilled its first offshore well in Prince Edward Island's Hillsborough Bay. The continental shelf off Nova Scotia was the site of important natural gas discoveries, and that province can boast oil production from Cohasset, Canada's first producing offshore oil field. Another offshore field, Hibernia, is in Newfoundland's Grand Banks. With field development costing $5.2 billion, that project may produce the world's costliest oil.

Twenty years after Alberta's transformation began, Canada's geographical hinterlands – the north and the offshore – also began to show tantalizing glimmers of petroleum riches.

Canadians have pioneered in the pipelining business for a century and a half. In 1858, a 25-kilometre pipeline (then the world's longest) began carrying natural gas to Trois-Riverès, Quebec. The world's first oil pipeline connected an oil field in Petrolia, Ontario to nearby Sarnia in 1867. Today, Canada is still home to the world's longest crude oil pipeline, and hosts the world's second longest natural gas pipeline. Nowhere does the industry use better or safer pipeline technology.

And nowhere have petroleum projects and policies been more politically explosive than in Canada. A political drama erupted in 1956, when a House of Commons debate on TransCanada PipeLines brought down the Liberal government of Louis St. Laurent. In 1980, energy issues which included an unpopular gasoline tax helped topple the weak Conservative government of Joe Clark. And the most contentious energy policy to date, the National Energy Program (NEP), created outrage when the Liberal Trudeau government introduced it later that same year.

The producing provinces perceived the NEP as encroaching on their political and financial turf. Memories of that trauma lingered. During the unsuccessful attempts at constitutional reform in 1992, Alberta insisted upon the creation of an equal, effective and elected Senate. It was hardly coincidence that the proposed new Senate would be able to veto legislation that changed tax policy on natural resources.

Petroleum represents Canada's biggest resource industry by a large margin. In sales, it is the fifth biggest non-financial industry, after food, consumer goods, motor vehicles and construction. Canada is the world's ninth largest producer of oil, the third largest producer of natural gas. Primarily because of sulphur stripped from gas, Canada is the foremost supplier of sulphur to international markets.[13]

Canada is unique among the world's advanced industrial countries in being a net exporter of oil, gas and all other energy commodities. And trade in energy makes up a big part of the national economy. During the five years ending in 1991, and despite the 1986 oil price collapse, petroleum generated a trade surplus of $26 billion. During the same period, Canada's total merchandise trade surplus was $15 billion. Put differently, oil and gas provided trade surplus and then some.

THE FUTURE OF OIL

What does the industry's future hold? The potential is unfathomable. Most of Canada's present oil-and-gas production comes from the Western Canada Basin, which stretches along the American border from Manitoba almost to British Columbia. From that base it narrows north by northwest to the Beaufort Sea. Originally, recoverable oil in that basin was about 19 billion barrels.[14]

Canada's three important frontier basins – the Beaufort Sea, the Arctic Islands and the East Coast offshore – have almost the same potential, and virtually none is on production. The potential for light and medium oil recovery in the frontiers is almost 17 billion barrels.[15]

Exploration in harsh frontier environments has made Canada the world leader in countless Arctic and offshore technologies and exploration techniques. In 1920, Norman Wells became the world's most northerly oil field,

THE GREAT OIL AGE   17

but Russian oil fields later assumed that distinction. The mantle of being the world's most northerly producer came back to Canada in the mid-1980s, however, when the small Bent Horn oil field in the Arctic Islands began producing.

That field is more than a symbol of Canada's sovereignty in the sparsely-populated north. Since the 1950s, Canada has used petroleum policy to assert territorial claim to those rugged, barren lands.

Although Canada's frontier and conventional oil resources are immense, they stand in the shadow of the non-conventional resources. Chief among these are the oil sands, the world's largest known oil deposits. The oil in those sands probably totals 2.5 trillion barrels. Of that almost incomprehensible volume, 33 billion barrels are recoverable by strip mining.

Unique in the world, the strip mining of oil has enabled Canada to become the undisputed world leader in synthetic oil production and technology. The Syncrude oil sands plant in northern Alberta takes hydrocarbon ore from the planet's biggest mine and processes it through one of the world's most complex industrial plants. That single plant produces 11 per cent of Canada's oil.

While the oil sands plants are a technological marvel, history has witnessed projects which were anything but. A startling example is a plan which came within a whisker's breadth of actual completion – a proposal to detonate a nine-kiloton nuclear device under an oil sand reservoir. In theory, the atomic blast would melt the thick, tarry oil so pumps could suck it to the surface. The plan died because of public unease about nuclear contamination.

However, in 1992 a provincial agency gave the idea new life. It proposed using controlled heat from a nuclear plant to produce oil sand reservoirs too deep for strip mining. And dramatic advances in underground mining techniques, pioneered in Alberta, have put 600 billion barrels of deeper oil within technological reach. That amount is roughly equal to one and a half times the reserves of the entire Middle East.

As if that resource bounty were not enough, the heavy oil belt along the Alberta/Saskatchewan border contains 37 billion barrels of oil.[16] Although heavy oil is expensive to produce, it is readily accessible. With conventional fields depleting, this resource has become a critical part of Canada's oil supply.

And then there is natural gas. Canada is the world's third largest natural gas producer, after the United States and Russia and the world leader in technologies that strip sulphur from sour gas. So sophisticated is this technology that the industry drills sulphur wells that produce almost pure hydrogen sulphide. Properly treated, this potentially deadly gas yields blocks of yellow sulphur.

There is enough natural gas in western Canada to supply our domestic and export markets well into the next century. And the likelihood of more discoveries means this premium, environmentally-attractive fuel can be plentiful through the lifetimes of almost every living Canadian.

As Canada goes about the business of developing native oil and gas potential, this country will probably construct more of the world's largest industrial projects. Arctic pipelines will cost as much as $10 billion. New oil sands plants will be so technically complex and physically huge that many will look upon $5 billion or $6 billion price tags as a bargain.

How much potential Canada can develop is a matter of conjecture, however. While the resources are vast, there are financial, technical and environmental limitations on development. Beginning in the latter 1980s, the petroleum industry underwent a structural shift so profound it brought many of the industry's core beliefs into question.

No longer was money available to keep the petroleum industry growing. A mature industry in a rapidly maturing Western Canada Basin found itself besieged with rising operating costs, lower oil and gas prices, increasing environmental responsibilities and competition from more attractive prospects elsewhere. These ingredients led to a drastic drop in drilling and development proposals. Nearly certain projects suddenly went on the shelf.

A cash-strapped industry experienced rapid and painful change. Companies took every conceivable step to cut costs. On the positive side of the ledger, these included technical and administrative innovations. On the negative side, bankruptcies and mergers became the order of the day.

In terms of human dislocation, there was no precedent in the industry's history. Almost every large corporation underwent brutal rounds of layoffs. A once rich, overmanaged and somewhat arrogant industry cultivated ever-leaner corporate cultures and ever-flatter management structures. Morale crept into the cellar. For an industry accustomed to growth, the times were unimaginably bleak.

As this volume goes to press, however, Canada's petroleum industry has weathered its most profound crisis and is beginning to mend. As the industry applied its talents and energies to struggles that often put corporate survival into play, the large oil companies came back – smaller and less influential but still strong. And in the downsizing of properties and personnel during the years of crisis, smaller companies found opportunities for growth. These years showed once again that Canada's petroleum industry is an enterprise steeped in drama and romance, fortune and folly. As these pages venture to show, that has been the case since the beginning of the Great Oil Age.

Chapter 2

The Source

Oil well site near Oil Springs, Ontario, 1972.
(Glenbow Archives/NA-2864-4179a)

The petroleum industry is a youngster. Born little more than 100 years ago, it sprang from the earliest oil wells in Canada and the United States. They tapped a resource endowment that took hundreds of millions of years for natural forces to form. For thousands of years before oil became an industry, people used petroleum for religious, domestic and commercial functions. This chapter examines petroleum from its geological origins to early records of its use by man.

WHAT IS PETROLEUM?

A range of substances help define the word petroleum. All are mixtures of mostly hydrocarbon, or molecular compounds of hydrogen and carbon. They form a spectrum of materials from light gases through liquids to heavy, gummy near-solids. The differences derive from varying proportions of hydrogen and carbon making up the petroleum molecule. At the light end, natural gas contains a high ratio of hydrogen to carbon atoms. At the heavy end, tarry bitumen contains a much lower hydrogen to carbon ratio.

Hydrogen and carbon can combine in an enormous number of ways. A variety of combinations are usually mixed together in a single reservoir. These petroleum mixtures also usually contain sulphur and traces of other elements and compounds. As a result, there are almost endless numbers of petroleum types, without clear-cut distinctions between them. But going from lighter to heavier hydrocarbons, petroleum generally falls into five categories: natural gas, natural gas liquids, light and medium crude oil, heavy oil and bitumen.

Origin of Petroleum

To understand the origin of petroleum, scientists have studied the geological history of the earth, considered astrophysical theories of the universe, and have examined the chemical reactions capable of generating the hydrocarbon chain. The many theories fall into two categories: inorganic and organic origins.

In the nineteenth century, the inorganic school of thought argued that unknown processes operating at great depth in the earth produced petroleum, and that its occurrence was associated with volcanic activity. Eugene Coste, known as the father of the Canadian natural gas industry, stoutly held this view. This theory was not quite satisfactory in its original form, because it left the actual process unidentified. Then, in 1866, French chemist Pierre Berthelot showed how hydrocarbons could be chemically derived by reacting carbonic acid with alkali metals. In 1877, the Russian chemist Dmitri Mendeleef – better known for his work with the periodic classification of the elements – suggested that percolating waters could react with iron carbide to produce hydrocarbons.[1] Another idea held that the early planet earth had an atmosphere of methane, and that the other hydrocarbons were derived from this original endowment.

Organic theories of petroleum origin are certainly older. An eighteenth-century French chemistry text attests to "the destruction of organic vegetable and animal products, buried in the earth and decomposed by the action of mineral acids" as the probable source of liquid and solid petroleum.[2] This approaches current theories. Others believed petroleum derived from underground deposits of coal. This idea too has carried forward to the present.

THE SOURCE

Those believing that plant and animal matter are the source material of petroleum have long argued over which of ancient history's plants and animals are responsible and whether plants or animals are the predominant source.

The debate between organic and inorganic theories of petroleum's origin has not ended. Inorganic theories are making a comeback.

According to conventional geology, oil and gas originated from plant and animal matter, buried in fine-grained sediments under oxygen-deficient conditions. Pressure from successive layers of sediment produced heat, and the combination of heat and pressure, combined with bacterial action and perhaps radiation, converted the organic matter into petroleum.

The petroleum then migrated along the paths of least resistance from the fine-grained rocks in which it originated. As it migrated, much of the petroleum passed through pores and cracks and unconsolidated material to the surface of the earth, and was thus lost to commerce. However, other hydrocarbons migrated into porous rock that was somehow closed off, forming the reservoirs that are the industry's bread and butter today.

Others advocate a new variation of inorganic theory they call abiogenic. The best-known modern advocate of inorganic origin is Thomas Gold, an astrophysicist at Cornell University. When explorations of space revealed that meteorites and other planets contained hydrocarbons in the total absence of life forms, it seemed to Gold to be strong evidence that petroleum could have originated abiogenically on earth as well.

Gold suggested that hydrocarbons may be abundant deep within planet Earth, and that the oil and gas already found originated at least in part in these deep zones. To test Gold's theories, a group drilled a deep well in the Siljan Ring, an impact crater in Sweden, and apparently found some 80 or more barrels of oil in a granite reservoir. But results proved inconclusive and the group began drilling a second well in 1991.

An iconoclastic Calgary geologist has developed even more radical theories than Gold's about the formation of oil and gas in the earth. Through a family-owned company, in 1991 Warren Hunt acquired the oil and gas rights to 960 000 hectares of Precambrian bedrock to test his theories. What is remarkable about this exploration play is that, according to conventional geology, he has acquired exploration rights in a geological region which could not possibly contain oil or gas.

In Hunt's two books, *Environment of Violence* and *Expanding Geospheres*, he proposes theories which, if proved, will fundamentally alter the geosciences. One test of his thinking is the exploration play in northern Alberta, which assumes that the Alberta oil sands had a deep-earth origin.

In Hunt's view, Earth's core contains vast amounts of hydrogen which can sometimes migrate toward the surface. Deep within Earth's mantle, it may react with silicon carbide to form gaseous hydrocarbons and silane gas. When disturbed, these brews move up to the underside of the earth's brittle granite crust. There, the silane can react with water to form silica sand. The slurry of sand, water and hydrocarbons is lighter than the granite above, creating instability.

Hunt believes the granite ruptured through what he refers to as the Carswell Gastrobleme, a 37-kilometre wide crater in northwestern Saskatchewan. Silica erupted violently, then oozed eastward from this conduit. Over time, 50 000 cubic kilometres of sand wound up sitting in a granite bowl across northwestern Saskatchewan – a phenomenon Hunt claims has never been explained geologically.[3]

He speculates that the shifting granite eventually resealed the Carswell rupture, trapping hydrocarbon-rich silica sand layers under Alberta's oil sands. His exploration play is based on the notion that only some of that oil rose to the surface to be degraded into today's oil sands. Hunt suggests that a great deal of conventional oil – perhaps hundreds of billions of barrels – could still be present in reservoirs west of the Carswell rupture, under the oil sands. If they exist, those reservoirs would have been formed by fractures in the granite which filled first with sand, then with abiogenic oil and gas.

Although it is a long-shot venture, Hunt has proposed a drilling program into granite which, if successful, will have a profound impact on conventional thinking about the origin of oil and gas. While the arguments continue in scientific circles, the fact remains that the large majority of geologists believe in some form of the organic theory of petroleum's origins.

THE ORIGIN OF PETROLEUM IN CANADA

Assuming the biological theory of petroleum, the origin of petroleum on this planet coincided with the creation of life. The span of time during which petroleum has been forming is difficult to comprehend. When the process began, even the most basic features of the earth's surface were drastically different from the ones we know today.

Until approximately 200 million years ago, the present continents of the earth were fused into a single gigantic continent called Pangaea. The part of Pangaea that became North America was subject to repeated ocean inundations and orogenies (periods of mountain building) over the millions of years before a period of uplift broke Pangaea apart and the modern continents began drifting away from mid-ocean ridges. Chains of enormous mountains were worn flat, again and again, by the forces of erosion. The resulting sediments were carried by wind and water to depressional areas – lakes, ocean margins, deltas – where they were laid down layer upon layer.

When life forms began to flourish in the oceans and then on land, their organic remains were deposited along with other sediments in a continuous rain on the lake and ocean floors. The pressure of the upper layers upon the lower ones compacted the lower layers into various kinds of stone: the mud into shales, the sands into sandstone, the shells of animals into the carbonates, limestone and dolomite. According to the most commonly-held theory, the fine-grained shales are the source rocks in which petroleum was born; the

THE SOURCE

coarser-grained sandstones and carbonates, which are more porous, became the reservoir rocks into which the petroleum would migrate and remain. A variation of this theory holds that limestone can be both a source and a reservoir rock.

Canada is well-endowed with sedimentary basins. Most of Canada's oil and gas production has come from the Western Canada Sedimentary Basin, a portion of North America's Great Interior Basin, which has its southern terminus in the Gulf of Mexico. The Western Canada Basin begins at the Beaufort Sea in the north, and stretches south through the Yukon and Northwest Territories. In southern Canada, it embraces the northeastern portion of British Columbia, almost all of Alberta, the southern half of Saskatchewan, and the southwestern corner of Manitoba. To the west, the basin is limited by the Cordillera, the series of mountain ranges which begins with the Rockies and continues west to the Pacific Ocean. The eastern boundary is the Precambrian Shield, the huge expanse of exposed granite that covers two-thirds of Canada. Precambrian rocks formed before life was plentiful on this planet, and so this enormous area is unlikely to contain petroleum.

The enormous periods of time shown by the earth's rocks can be seen on the geological time scale chart on page v in the front matter. The majority of the oil reserves in the Western Canada Basin are found in reservoirs of Devonian Age,[4] rocks that date back as much as 390 million years.

The limestone remains of coral reefs formed in the ancient inland seas have proved a particularly fruitful hunting ground for oil in Alberta. By contrast, about half of the gas reserves of the Western Canada Basin are found in Cretaceous reservoirs (rocks 140 million to 67 million years old).[5] Gas found in older Paleozoic reservoirs frequently contains hydrogen sulphide, and is known in the industry as "sour gas."

Although the Western Canada Basin is the most heavily explored and the most prolific of Canada's major sedimentary basins, the first explored was the basin in southern Ontario. This basin yielded several small oil and gas finds. It is one of several, all of Paleozoic age, found in eastern Canada. The largest and least explored of these is the Hudson Bay Basin.

In several areas of Canada's far north, great depths of sediment accumulated and enclosed a bounty of organic matter and potential oil or gas reservoirs. The basin areas can be divided into those associated with the Beaufort Sea and the Mackenzie River delta, and those associated with the Arctic Islands. The Beaufort-Mackenzie Delta area is a continental shelf where great thicknesses of deltaic sandstones and shale overlie Paleozoic rocks. Exploration of this region has only really begun, and proceeds slowly because of difficult climatic, logistical and economic conditions. Nonetheless, the petroleum industry made important discoveries of both oil and gas there in recent decades.

Even less well explored are the basins of the Arctic Islands. Exploration in the thick sediments in these areas has produced large gas discoveries and significant oil pools. Development is hampered by the shifting Arctic ice pack under which many petroleum prospects lie.

The west coast of Canada and the inter-mountain areas of British Columbia include several basins which contain hydrocarbons. But these basins have not yet yielded significant discoveries or production.

The east coast offshore is proving more prolific. Sedimentary basins along the continental margin stretch for 5500 kilometres, most of them the result of sedimentary deposition after the break-up of the continents. Exploration in some of these basins found several important discoveries. These include the Hibernia oil field on the Grand Banks of Newfoundland, and the gas fields surrounding Sable Island on the Scotian Shelf. Optimists expect production from the east coast offshore in the late 1990s.

While this review of Canada's sedimentary basins indicates that important discoveries have been made in the geological frontiers that is, the northern and offshore sedimentary basins, it is important to note that most of these discoveries are only indications of the geological potential of those areas. A basin's geological significance and its commercial significance are related, but not the same.

A geologically important find, for example, one of the discoveries in the Beaufort Sea may include a major accumulation of oil. But the costs associated with developing the field and delivering the oil to market may not justify the price it will fetch at the refinery gate. In that case, the field would be a geological success but a commercial failure. By contrast, a virtually identical field located in Alberta could be a commercial bonanza, since it could probably be easily and economically developed. The Beaufort Sea reservoir used in the illustration could only be developed if oil prices reached a high enough level to justify development costs.

The following table summarizes the resource potential of Canada's major sedimentary basins, but without reference to their economic potential, since the commercial value of most frontier discoveries is still in doubt. Only time will disclose how much of Canada's oil and gas resource potential will ever be actually produced.

PETROLEUM USE

Once it is squeezed from its source rocks, petroleum migrates from high pressure to lower pressure spaces within rocks. In general, this means it migrates upwards. The oil or gas usually meets an impermeable layer of rock. If not, it continues its forced march upward to the surface of the earth. Thus tar, oil and gas seeped onto the earth's surface for millions of years.

With the evolution of people into a thinking, material-using species, one of the cornucopia of materials available for exploitation was the petroleum at surface seeps. Although we do not know how people used petroleum prior to recorded history, paleolithic peoples, like North American Indians, apparently used it for medicinal purposes.[6] Excavations of some of the earliest sites of civilization show signs of petroleum use. In the Euphrates River valley, the ancient civilization of Ur used bitumen as a brick mortar. The people of this society also used bitumen as a kind of glue for holding flint cutting edges in bone sickles.[7] In the Bible, Noah's instructions for the building of an ark included the command to "pitch it within and without with pitch" (Genesis 6:14). The ancient civilization of the Indus Valley used bitumen to waterproof its baths as long as 5 000 years ago.[8]

Gas seeping out of the earth, ignited by lightning, could burn for as long as the gas flowed. Such "eternal fires" were objects of veneration for many early religions. Other ancient records claim that barbarians carried naphtha with them and would inspire fear and awe by igniting it in sight of the homes of local leaders.[9]

The Chinese civilization was probably the first to harness and use natural gas. The Shu Han Dynasty used natural gas 200 years B.C. to light its temples and to evaporate brine from salt wells. By the tenth century A.D. the Chinese of Peking succeeded in transporting natural gas by bamboo pipeline and used it for street lighting.

The ancient languages used in Greece and Rome gave us basis for the word "petroleum": "petra" is Greek for rock; "oleum" is Latin for oil. One of the roles played by petroleum in ancient Greece was to supply the Oracle of Delphi with some of its mystical authority from a burning gas seep.

Toward the end of the Middle Ages, petroleum began to serve a variety of purposes in Europe. Europeans used it for medicine, paving, lighting and caulking ships. Some primitive refining methods were applied to raw petroleum. A pamphlet published by Johan Volck in Strassbourg in 1625 explained petroleum refining methods of the day. Uses included lamp fuel, dressings for wounds, wood preservative, and lubricant.[10]

Although refined oil provided illumination for many Europeans and North Americans in the nineteenth century, it was not the only petroleum-based fuel. Gas derived from coal-lit lamps in London streets in 1807, and coal gas streetlighting made its Canadian debut in Montreal in 1836.

The Chinese precedent of using natural gas for this purpose was not followed again until 1821 when the residents of Fredonia, New York, brought natural gas into town from a shallow nearby well by a hollow log pipeline, and lit their streets with it. Fredonia called itself "the best-lit city in the world." Because of severe leakage problems, other towns refrained from trying to become as well-lit.

A common eighteenth and nineteenth century source of lubricant and illuminating fuel came from the whaling industry. Whale oil supplied much of the household demand for lamp oil. As whalers pursued their prey to the far corners of the earth, they supplied ever larger markets for lamp fuel and lubricants. But whales could not have continued to meet the growing demand, especially as hunting depleted their numbers. Society needed a more secure and less costly source of fuel and lubricant.

Since the amount of petroleum found in seeps and gum beds was insufficient to fill the gap, clever inventors set about distilling usable oils from oil shales and coal. This approach gained ground swiftly when the Williams well in Ontario and the Drake well in Pennsylvania made underground supplies of oil plentiful for the first time. The industries devoted to lubrication and illumination swiftly shifted to this cheap and plentiful commodity.

Chapter 3

The Early Years

Canadian type pole-rig drilling for Tertiary coal seams on Souris River near Roche Percee by Geological Survey of Canada, 1880. (Glenbow Archives/NA302-10)

    A thirsty army of new machines powered the Industrial Revolution and by 1850, their demand for fuel and lubricants exceeded conventional supplies. Household and commercial use of illuminating fuels also increased. As whale populations declined, the price of whale oil rose quickly. Efforts to distil illuminating fuels from oil shales and bitumen deposits, though making progress, failed to meet the increasing demand. And smelly supplies of coal gas, although available in some cities, were not a long-term solution for urban lighting and fuel.

It was only a matter of time before shovels began searching for oil. This in itself was not a radical step. Evidence suggests that the Chinese dug for oil 2 000 years ago. Throughout history the quest for water and for salt had occasionally run into nuisance deposits of oil, which foiled and fouled these efforts. By the mid-nineteenth century, however, the industrial climate was ripe for oil to become an important commercial enterprise.

Those who eventually dug the epoch-making oil wells of the late 1850s benefitted from technological innovations pioneered by the Ruffner brothers, David and Joseph. When they sank a well in a West Virginia salt lick in 1806 they fought to keep mud and gravel from caving into their hole. Their solution was a hollow sycamore tree. As they dug, the sycamore slid down, protecting the sides of the hole.

When the Ruffners arrived at bedrock and still did not have the quantity of brine they wanted, they went deeper, using a steel bit hung from a long iron rod which they repeatedly dropped to smash the stone. To get the springing action needed for repeated strikes, they devised a spring pole (probably the green trunk of a young tree). The driller used his weight to drive the drill bit down, then let the spring in the tree bring it back up. Though the Ruffners can not take total credit for this development, since the Chinese had been doing it for thousands of years, they did make a few improvements that qualify as invention. The Ruffners thus penetrated bedrock and got their brine. To raise the brine to the surface undiluted, they made one more innovation. Constructing a pipe from long strips of wood, they pushed it to the bottom of the hole. Brine drawn through this tube did not touch the water that stood in the hole above the brine. By the time the Ruffners completed their well in 1808, they had invented casing, cable tool drilling and tubing.[1]

THE FIRST OIL WELLS

With oil scarce and fetching a high price, and with the Ruffner technology providing a method for penetrating both soil and bedrock, it was only a matter of time before entrepreneurs began drilling for oil.

The first recovery of petroleum for commercial use in Canada was by Charles Nelson Tripp in Enniskillen Township on the north shore of Lake Erie. Tripp's dabbling in the mysterious "gum beds" of that locale near Black Creek in 1851 led to the incorporation of Canada's first oil company. Parliament chartered the International Mining and Manufacturing Company in 1854, with Tripp as president. The charter empowered the company to explore for asphalt beds and oil and salt springs, and to manufacture oils, naphtha paints, burning fluids, varnishes and related products.

Although Tripp's asphalt received an honorable mention at the Paris Universal Exhibition in 1855, financial success eluded his company. Several factors contributed to the downfall of the operation, including the lack of roads which made the movement of machinery and the distribution of the products difficult. After every heavy rain, the area became a swamp.

A carriage builder named James Miller Williams became interested and visited the site in 1856. Tripp unloaded his hopes and properties on Williams, reserving a spot for himself on the payroll as landman. Williams incorporated J.M. Williams and Company in 1857 to exploit the Tripp properties. Stagnant, algae-ridden water lay almost everywhere and, looking for better drinking water, Williams dug a well a few yards down an incline from his asphalt plant. At a depth of 20 metres, the well struck free oil instead of water. In 1858 it became the first oil well in North America, remembered as Williams No. 1 at Oil Springs, Ontario.

The famous Edwin Drake discovery well went into production on August 28, 1859 near Titusville, Pennsylvania. It vies for the claim as North America's first oil well. Admittedly the Drake well was no water well gone wrong, but a true oil well from start to finish. Backed by the Seneca Oil Company, the well drilled among the oil springs beside Oil Creek, and poked down into bedrock a total of 21 metres before filling with oil. The supporters of Colonel Drake's claim to the first oil well in North America point out that his well found oil beneath the bedrock, while the Williams well in Ontario found oil above. Both wells stand as milestones in the history of the North American oil industry. Each well touched off a flurry of events, the oily product of which eased the plight of industrialists and householders desperate for lubricants and fuel.

Williams eventually abandoned his Oil Springs refinery and transferred his operations to Hamilton. In 1860, the local newspaper carried his ad: "Coal oil for sale, 16 cents per gallon for quantities from 4 000 to 100 000 gallons." Williams reincorporated his firm as The Canadian Oil Company and operated facilities for petroleum production, refining and marketing – a mix that qualifies his company as the first integrated oil company in the world.

Exploration in Lambton County quickened with the discovery of free flowing oil in 1860. Until then, hand pumps coaxed oil from the ground. But the first gusher blew in on February 19, 1862, when Hugh Nixon Shaw struck oil at 48 metres. For a week the oil gushed unchecked, coating the distant waters of Lake St. Clair with black film. Dr. A. Winchell, in his *Sketches of Creation*, refers to the event in these words:

> Though Western Pennsylvania has produced many flowing wells of wonderful capacity, there is no quarter of the world where production has attained such prodigious dimensions as in 1862 upon Oil Creek in the Township of Enniskillen, Ontario. The first flowing well was struck there January 11, 1862, and before October not less than 35 wells had commenced to drain a storehouse which provident nature had occupied untold thousands of

years in filling for the uses of man. The price had fallen to ten cents a barrel, three years later that oil would have brought ten dollars a barrel in gold. From detailed determinations I have ascertained that during the spring and summer of 1862, no less than five million barrels of oil floated off upon the waters of Black Creek.

Within a few years the wells were producing mostly salt water and the boom moved eight kilometres north, to Petrolia. Crews drilled ten thousand wells there, a rail line replaced the oxen trails and a pipeline carried oil to the refineries and docks at Sarnia.[2]

Although the industry in Central Canada began with a promising start, Ontario's status as an important oil producer declined rapidly. Canada became a net importer of oil during the 1880s, and dependence on neighboring Ohio for crude oil increased following the arrival of Canada's first automobile in 1898.

TECHNOLOGICAL ADVANCES IN THE ART OF DRILLING

While Williams brought in Canada's first well with a shovel and backbreaking labor, the famous American discovery used only slightly less primitive cable-tool technology. Drake's rig also used steam power. Early Canadian rigs relied on foot power and the recoil of the spring pole. Steam-powered rigs arrived from the United States shortly after 1860. In the early American rigs, the string of heavy tools and the chisel-edged bit hung from a manila cable – hence the phrase "cable tool drilling." In Canada, in the 1860s or seventies, a rig emerged which suspended the drilling tools from a series of linked hardwood rods. The pole-tool rig also originated in the United States, but because Canadian drillers preferred it to the cable tool rig, it became known as the Canadian rig.

Both rigs used a walking beam which rocked over a fulcrum called the "samson post," pulling and dropping the tools and bit in the bottom of the hole. A bailer replaced the drilling tools and cleaned cuttings from the bottom of the hole. Canadian entrepreneur W.H. MacGarvey helped develop the Canadian rig. He also made Canadian drilling technology and the Canadian driller famous around the world.

MacGarvey used Canadian rigs to develop an oil field in Austrian Galicia. Soon, the Canadian rig became the preferred drilling device in central Europe.[3] Although few records remain, it is certain that drillers from Petrolia worked in Java, Peru, Turkey, Egypt, Russia, Venezuela, Persia, Baluchistan, Rumania, Austria and Germany. MacGarvey amassed a large fortune from his efforts but saw his Galician properties destroyed when the First World War swept across Europe.

Another drilling method, making a hole by rotating a sharp bit, is at least as old as the idea of pulverizing the earth with repeated blows. Egyptians drilled holes using the rotary method in stone quarries as early as 3000 B.C.[4]

Leonardo da Vinci left sketches of a rotary auger drawn circa 1500.[5] The French used dry rotary drilling to procure well water during the seventeenth and eighteenth centuries and, in 1844, Robert Beart of England received a patent for a fluid circulating system for rotary drilling.

Drillers used a crude version of the rotary rig in the 19th century petroleum industry, often powered by a mule walking in a circle.[6] But the best known pioneering effort in rotary drilling took place in 1901 with the drilling of the Spindletop well near Beaumont, Texas. Besides being the first big success in the oil patch with rotary equipment, Spindletop was also a legendary American gusher. It proved the existence of petroleum in salt domes and it was the first rotary well to use drilling mud as a circulation medium.[7]

Rotary drilling equipment invaded the Turner Valley oil field in southwestern Alberta during the 1920s, and entirely displaced cable-tool drilling by the late 1930s. The first rotary equipment in Turner Valley used steam derived from coal. Rigs eventually used natural gas as it became available. As the transition from cable tool to rotary rigs progressed, some rigs used both rotary and cable tool machinery. These "combination rigs" used either method of drilling, thus taking advantage of the appropriate technology.[8]

Derrick floor, running drill pipe into hole, north end of Turner Valley, 1942. (Provincial Archives of Alberta, H. Pollard Collection, P1341)

BIRTH OF THE NATURAL GAS INDUSTRY IN CANADA

Canadians knew about natural gas long before they put it to any practical use. Around 1820, youngsters of Lake Ainslie, Nova Scotia, amused themselves by driving stakes into the ground, removing them, then firing the escaping natural gas. In 1859, Dr. H.C. Tweedle found both oil and gas in what became the Dover field near Moncton, New Brunswick, but water seepage prevented production from these wells. Tainted with foul-smelling and toxic hydrogen sulphide, sour gas greeted drillers in 1866 near Port Colborne, Ontario, during the oil drilling boom. It was a harbinger of the gas fields found later in the southwestern part of the province.

Eugene Coste, the Paris-educated geologist from Ontario (mentioned earlier as a stout advocate of the inorganic theory of the origin of petroleum), brought in the first producing gas well in Essex County, Ontario, in 1889. His work in that area eventually led to the formation of the Ontario utility, Union Gas.

Canada first exported natural gas in 1891 to Buffalo, N.Y. from the Bertie-Humberstone field in Welland, Ontario. A 20-centimetre pipeline under the Detroit River eventually transported gas from the Essex field to Detroit. By 1897, a pipeline to Toledo, Ohio taxed the Essex gas field to its limits. As a result, the Ontario government revoked the pipeline licence and passed a law prohibiting the export of natural gas and electricity.

The central Canada gas industry reached another important milestone in 1911 when a merger of three companies using Ontario's Tilbury gas field formed Union Gas Company of Canada Limited.

In 1909, New Brunswick's first successful gas well came in at Stoney Creek near Moncton. This field still supplies customers in Moncton and has produced nearly 800 million cubic metres of gas and 130 000 cubic metres of oil. Propane now supplements the limited supply of natural gas from the field to meet the city's needs.

In western Canada, petroleum discoveries date back to 1883. Natural gas greeted water well drillers at the Canadian Pacific Railway Siding No. 8 near Langevin, west of the current city of Medicine Hat, Alberta. An accidental find, the gas discovery shocked the drillers who sought a reliable supply of water for the CPR's steam-driven locomotives. The gas flow caught fire and destroyed the derrick.

This find prompted Dr. George Dawson of the Geological Survey of Canada to make a notable prediction. Having observed that the strata penetrated in this well were continuous over great areas of western Canada, he prophesied that the territory would some day produce large volumes of natural gas.

A well drilled near Medicine Hat in 1890, this time in search of coal, also encountered a large flow of natural gas. The find prompted town officials to approach the CPR with a view to drilling deeper wells for gas.

The resulting enterprise led to the discovery of the Medicine Hat gas field in 1904. The community took advantage of the natural resource and became the first urban area with a gas utility in western Canada.

In 1894, the Dominion Government brought a rig from Toronto to northern Alberta to drill for oil along the Athabasca River. The second well drilled by this rig at Pelican Rapids struck gas in 1897 and then blew wild. Out of control, it burned for 21 years, consuming natural gas at estimated 240 000 cubic metres per day until 1918.

Natural gas service began in Calgary at the beginning of this century when A.W. Dingman of Toronto formed the Calgary Natural Gas Company. He drilled a successful well in east Calgary, laid pipe to the Calgary Brewing and Malting Company site and provided gas to the brewery on April 10, 1910. Gas mains soon provided domestic fuel and streetlighting.

In another development, Eugene Coste moved west and drilled the locally famous Old Glory gas well near Bow Island, Alberta, in 1909. In 1912, his Canadian Western Natural Gas Company built a 280-kilometre pipeline connecting his Bow Island field to Lethbridge and Calgary. It augmented the Dingman enterprise in Calgary, which was unable to supply the growing demands of the city. By 1913, several other towns in southern Alberta also boasted natural gas service from the Canadian Western system and Coste's pioneering enterprise provided fuel to nearly 7 000 customers.

OIL IN WESTERN CANADA

John George (Kootenai) Brown was probably the first man to attempt to develop western Canada's petroleum potential. An Irish frontiersman with an Eton and Oxford education, he was also an early homesteader in this region. In 1874, Brown filed the following affidavit with Donald Thompson, the resident solicitor at Pincher Creek:

> *[I was] engaged as a guide and packer by the eminent geologist Dr. George M. Dawson, and he asked me if I had seen oil seepages in that area, and if I did see them, would I be able to recognize them. He then went into a learned discussion on the subject of petroleum. Subsequently some Stoney Indians came to my camp and I mixed up some molasses and coal oil and gave it to them to drink, and told them if they found anything that tasted or smelled like that to let me know. Sometime afterward they came back and told me about the seepages at Cameron Brook.*[9]

In 1901, John Lineham of Okotoks, Alberta, organized the Rocky Mountain Drilling Company and in 1902 drilled the first exploration well in Alberta on the site of these seepages. Now part of Waterton Lakes National Park, the Historic Sites and Monuments marker commemorates the discovery well and Oil City, the boom town which sprang up briefly in the area. The discovery well briefly produced up to 350 barrels of oil per day, but neither this well nor seven later exploration attempts resulted in steady production. Perhaps the greatest contribution of the Oil City play came about when the Western Oil and Coal Company drilled there and collected 256 rock samples at different depths which they examined for traces of oil. This method of systematic sampling set a precedent that drillers now routinely follow.[10]

THE END OF AN ERA

As World War I approached, the oil industry in central Canada waned while the petroleum potential in the west struggled to become commercially viable. As southern Ontario oil and gas fields declined, the word "inexhaustible" disappeared from news reports about them. Sages recommended that entrepreneurs turn their energies to manufacturing gas from coal, because a growing nation could not rely on petroleum. In Alberta, the natural gas industry blossomed quickly and, for a few years, had the reserves to support local markets. But the oil industry in the west showed no signs of bursting into production. Despite endless tantalizing seeps and shows, the wildcat wells disappointed speculators and observers. Most wells ended up dry and abandoned. George M. Dawson's 1888 map of the petroliferous areas of western Canada seemed to promise petroleum wealth. But where was it? Where should one look?

Chapter 4

Oil Between the Wars

Flare at Turner Valley well, 1930s.
(Glenbow Archvies/NA-67-143)

For many years Indians, ranchers and then settlers frequented the quiet foothills of the Rocky Mountains southwest of Calgary. Then a rancher struck oil and for more than 30 years western Canada's petroleum industry focused squarely on a curious valley. Named after two brothers who began ranching there in 1886, Turner Valley sat atop an enormous geological structure which contained naphtha-soaked, or wet, natural gas along two horizons and oil in a deeper reservoir. Gas and oil from Turner Valley fuelled urban and industrial development in Alberta and, during World War II, fuelled the planes of a Commonwealth air training program. For some time it seemed that Turner Valley might be the only major petroleum field in Canada. That prediction seems ridiculous in retrospect, but it was the premise upon which oil companies and governments relied until the discovery of oil at Leduc in 1947.

The Turner Valley era is in fact three periods characterized by the three exploration discoveries that instantly and substantially changed the industry's perception of the field. These periods were the Dingman era, the Royalite #4 era and the Oil Column era.

THE DINGMAN ERA

In early 1914, oil fever swept Calgary. Investors lined up outside makeshift brokerage houses to get in on exploration activity triggered by a wet gas discovery at Turner Valley.[1] So great was the excitement that in one 24-hour period promoters formed more than 500 "oil companies." Although incorporated in 1913, the Calgary Stock Exchange was unable to control the unscrupulous practices that relieved many Albertans of their savings.

Calgary Petroleum Products drilled Dingman #1, the well behind this speculative flurry, near the crest of the great structure that underlies Turner Valley.[1] Using an American style cable tool drilling rig, the tools hung from the longest manila drilling line ever used. Spudded in January 1913, the well came in with a roar on May 14, 1914.[2] It found the reservoir at 664 metres and soon produced 4 million cubic feet of gas per day from the Cretaceous sandstone horizon.[3] The gas dripped with smelly naphtha, a light oil condensate, pure enough to burn in automobiles without further refining. Fame and success greeted Bill Herron, William Elder and driller Archibald Dingman as well as the other partners in the syndicate that created the Calgary Petroleum Products Company.

While the Dingman well and its successors established the first commercial field in western Canada, the high expectations raised by the discovery did not last. The few wells from the ensuing boom that struck gas in the Cretaceous sandstone produced only small volumes of naphtha. By 1917, the Calgary City Directory listed only 21 "oil mining companies," compared to 226 in 1914.

THE ROYALITE #4 ERA

Financial hardship followed the Calgary Petroleum Products Company (CPP). Finally, the small processing plant attached to its Dingman #1 and #2 wells burned down in 1920. Royalite Oil Company, a new Imperial Oil subsidiary, arose from the ashes of the fire, taking over the CPP interests. In late 1921, Canadian Western Natural Gas began buying processed Turner

Valley gas from Royalite. The new supply arrived just in time to help meet the growing demand of consumers who were using more gas than the rapidly depleting Bow island gas field could produce.

Calgary Petroleum Products Company refinery at Turner Valley before 1920. (Provincial Archives of Alberta/PAA-85.248 Box 30)

Taking up where its predecessor had left off, Royalite began a drilling program which included the legendary Royalite #4 well. Spudded in 1923, this well became the first to drill test the Paleozoic rocks, at a much deeper level than the formation that produced the naphtha in the Dingman wells. It did not start out that way; in fact it blew in the first time at the relatively shallow depth of 875 metres. This was above the productive sandstone layer of the earlier wells and, when #4 started to decline sharply, Royalite drilled deeper into the sandstone horizon. The well pierced the sandstone and struck limestone without any further discovery. The drillers pushed on and, at a depth of 1 140 metres, the drilling tools stuck in the hole. During attempts to fish the tools out, the well blew in at 21 million cubic feet of gas per day.

The gas contained high volumes of naphtha and a high percentage of hydrogen sulphide. Since hydrogen sulphide is a highly poisonous gas, Royalite built a sweetening plant to take the dangerous substance out before it sold it to Canadian Western.

Trouble began at Royalite #4 when the crew attempted to clamp a valve on the well to shut it in. The shut-in pressure rose at alarming speed. When the gauge hit 1 150 pounds per square inch, the drillers ran for their lives. In 20 minutes, 939 metres of 21-centimetre pipe and 1 052 metres of 16-centimetre pipe, a total weight of 85 tonnes, rose out of the ground to the top of the derrick and the well blew out again.

On October 19, 1924, the gas caught fire. Residents of Lethbridge, more than 150 kilometres away, saw the light from the burning well. The fire blazed for 21 days. When the combined flow of steam from seven steam boilers failed to snuff out the flame, a charge of dynamite extinguished the blaze. It took Oklahoma wild-well experts two more months to bring Royalite #4 under control.

The reaction to this discovery was quite different from the sudden boom that had followed the first Dingman well. The next few wells drilled into the Paleozoic limestone did not produce similar results, and many explorers took it to mean that the productive area was small.[4] It was five years before a real boom began. Home Oil's discoveries in a sandstone reservoir, when deepened, produced from the limestone horizon and established the field. Activity slowed during the Depression, but about 100 naphtha wells met with success in the decade following Royalite #4.[5]

THE OIL COLUMN ERA

The wells of both the Dingman and the Royalite #4 eras at Turner Valley produced only naphtha, not crude oil. Similar to gasoline, naphtha served as fuel for local automobiles and tractors. But the process of stripping naphtha from the gas also produced huge volumes of natural gas for which there was no market. Since Royalite had the only natural gas pipeline to Calgary consumers, the remaining Turner Valley producers flared the byproduct gas after producing the valuable liquid product. Also, with the Royalite #4 blowout fresh in their minds, fear of a similar accident prompted most producers in Turner Valley to just flare excess gas. Day and night, the huge flares burned enough fuel to heat thousands of homes.

Royalite burned off much of the surplus gas in a small ravine known to locals as "Hell's Half-Acre." Grass beneath the flares stayed green year-round and local hunters stalked game by the artificial light. Depression hobos slept in the circle of warmth. Reporters from around the world described the flares in florid terms. One Manchester scribe wrote that,

> *Seeing it, you can imagine what Dante's inferno is like . . . a rushing torrent of flame, shooting forty feet high . . . a ruddy glow to be seen for fifty miles . . . . Men have seen the hosts of hell rising, the titanic monster glowering from the depths of Hades.*[6]

This was waste by anyone's definition, but the producers cared little for a product for which they had no market. When the Alberta government attempted to control the waste in 1932, producers appealed to the Supreme Court of Canada. It declared the conservation orders *ultra vires*, or beyond the power of the province.

But waste of crude oil was a different matter. And on June 16, 1936, the day Bob Brown's Turner Valley Royalties #1 well began flowing 850 barrels of crude oil per day,[7] the industry realized it had been doing just that – wasting crude oil right along with the gas since 1914.

The Brown discovery meant that the great Turner Valley field was both a gas and an oil reservoir, with the gas cap surmounting and putting pressure on the oil. By producing and burning off billions of cubic feet of natural gas from that gas cap, the naphtha producers had decreased the natural gas pressure, or drive, on the oil reservoir. As a result, the Turner Valley field realized only a portion of its original oil potential. In 1938, the Alberta government finally created the Oil and Gas Resources Conservation Board. With a mandate to enact regulations to control gas production and make best use of the remaining gas cap, it did so. But the worst of the damage had already been done.

Although the gas waste reduced the amount of recoverable oil from the field, Turner Valley was still western Canada's first major discovery. Until the late 1940s, Calgary oilmen also boasted it was the largest producing oil field in the British Empire. Turner Valley production peaked during World War II when the federal oil controller doubled production quotas from the Turner Valley wells. During 1942, the field produced 10 million barrels of oil – more than 27 000 barrels of oil per day. Total wartime production for the field accounted for 95 percent of total Canadian output.[8]

TURNER VALLEY FIRSTS

Producers faced many technological problems in Turner Valley, and a few of those problems were new. Some drilling holes wandered as much as 22° off course. High-pressure gas caused freezing as the product expanded quickly. Gas hydrates blocked pipelines. Hydrogen sulphide threatened the lives of workers. High-pressure sour gas, casing failures, sulphide stress corrosion cracking, corrosion inside oil storage tanks, and the cold climate caused a host of other problems.[9]

In dealing with such problems, Turner Valley operators acquired new expertise and made technological improvements that helped earn the field its place in history.

Following are a few Turner Valley firsts – not firsts for the world, but milestones in the development of the industry in Canada.

Rotary drilling, introduced to Turner Valley in 1925, eventually replaced cable tool drilling everywhere in western Canada. Because the steam-driven rigs of that period delivered smooth power at relatively low operating costs, they became common in the valley until the early 1950s even though they were rapidly becoming an anachronism elsewhere. At the beginning of the modern era, this dated technology proved to be absolutely essential to bring the legendary Atlantic #3 blowout of 1948 under control.[10]

Another first for Turner Valley was the first true gas processing plant in Canada, constructed in 1914. The second Canadian hydrogen sulphide removal or sweetening facility arose there in 1925. The McLeod #2 well in 1927 introduced nitro-shooting to Canada. This process involved detonating an explosive charge within a well bore to increase the well's productivity. The first use of acidizing, a process that uses acids to etch small channels in the producing formation, thereby allowing oil and gas to flow more easily, helped extend the life of the Model #3 well in 1936. Gas storage began in 1944 and water injection in 1948.

A dispute between the federal government and the petroleum industry developed in response to the success in 1936 of the "royalties" system of financing wells. Under this system, pioneered on the Turner Valley Royalties #1 well, investors received a royalty-like percentage of total oil and gas production from a successful well. If a well was prolific, the return on investment could be high. This system was so successful that investors drilled 69 royalties-type wells in Turner Valley in the two years following the Royalties discovery. Since only two of those wells were dry, the primary constraint on new investment was the rapid saturation of local oil markets which accompanied the new production.

But in 1938, the federal government decreed that income from oil production was taxable as profits in the hands of the producing company. In the investor's hands it was taxed again as income, rather than as the return of capital. Although a producing company appealed this decision successfully, the incident shook confidence in the royalties system of finance. Indeed, in 1942 the government amended the Income Tax Act to tax oil income from royalty trusts at wartime rates. Although the federal government repealed this tax provision in 1950, the royalties system of financing never returned.

Also during the Second World War, Turner Valley saw the formation of Canada's first Crown oil company. Anxious to increase production, the Canadian federal government created Wartime Oils Ltd. to finance drilling in the Turner Valley field. Under the scheme, a leaseholder only had to repay drilling costs out of production from successful wells.[11] During a debate in Parliament, the minister in charge of wartime production, C.D. Howe, explained that the purpose of the company was not "to wildcat for oil." Howe said:

> It will drill out a particular area under a particular plan. To set up a government company which would provide money for anyone who wished to drill in wildcat areas would call for unlimited funds and the likelihood of obtaining returns would be exceedingly small.[12]

NORMAN WELLS

Turner Valley made most of Canada's petroleum industry news during the 1914-1946 period, but not all of it. Another exciting saga played out thousands of kilometres north at Norman Wells in the Northwest Territories.

The Norman Wells story goes back two centuries. In 1789, when Alexander Mackenzie was exploring the Mackenzie River in hopes of reaching the Pacific Ocean, he made a note in his journal about oil seeps issuing from the banks of the river. A century later, R.G. McConnell, exploring on behalf of the Geological Survey of Canada, made the same observation. In 1911 Jim Cornwall, a northern businessman, saw oil on the Mackenzie and hired a local Indian named Karkesee to look for seepages.

Discovery well, Imperial Oil Fort Norman No. 1, Mackenzie River, Northwest Territories, 1920.
(Glenbow Archives/NA-781-14)

Karkesee found several, and analysis of this oil found it similar to the crude oil from Pennsylvania. Cornwall formed a syndicate with two Calgary businessmen and the group engaged T.O. Bosworth, a prominent petroleum geologist, to study the area. During this 1914 study, Bosworth staked three claims on behalf of his backers and reported enthusiastically on the prospects of the area.[13] The outbreak of World War I put a halt to plans for development. By the end of the war, Imperial Oil owned the Bosworth claims.

In 1919, Imperial began exploratory drilling on the Mackenzie. Two wells in the Great Slave Lake vicinity found salt water. Farther down the Mackenzie, in the Fort Norman area, the third well showed oil.

Led by Ted Link, who later became Imperial's chief geologist, the crew that drilled the successful well consisted of several drillers, an ox named Old Nig, and a cable-tool rig. This crew made the trip from Edmonton in six weeks, arriving at the drill site on Bosworth Creek with just enough time to set up camp and erect a rig before winter set in. Legend has it that Ted Link chose the site by waving his arm and saying, "Drill anywhere around here."

The crew that wintered with the well ate Old Nig before the relief crew arrived the next July. During August, 1920, at a depth of about 1 240 metres, the world's most northerly oil well came in. The well produced 600 to 900 barrels per day at first, but settled down to an average of about 100 barrels daily.[14]

The geology of this discovery is unusual. Sandstone reservoirs were the best understood at the time and most believed that the oil at Norman Wells came from the sandstone horizon. In fact, the drill passed through the sandstone with no result and found oil in hard fractured shales below. The shale itself seemed too compact to contain oil in such quantity, so at first Imperial believed the oil resided in fractures in the shale. If true, it represented something new in petroleum geology. Subsequent drilling disproved this notion. The shale enveloped a coral reef of middle Devonian age, the same kind of fossil structure in which Leduc's oil appeared later. The oil resided in the reef. The discovery well at Norman Wells found a quantity of oil escaping through cracks to the surface.[15]

In the months following its discovery, Imperial drilled three more holes, two successful and one dry. The company also installed enough refining equipment to produce fuel oil for use by the local missions and boats. The refinery and oil field both closed in 1921, after the operation proved too costly to maintain.

Norman Wells marked another important milestone in 1921 when Imperial flew two all-metal, 185-horsepower Junkers airplanes to the site. These aircraft were among the first of the legendary bush planes which helped develop the north, the forerunners of today's northern commercial air transport. A new refinery opened at Norman Wells in 1936 to supply the Eldorado Gold Mines at Great Bear Lake, but the site did not become significant until after the United States entered World War II.

When Japan captured two of the Aleutian Islands, Americans grew concerned about the safety of their oil-tanker routes to Alaska. They began looking for an inland supply of oil, safe from attack. The United States and Canada negotiated to build a refinery at Whitehorse, Yukon, with crude oil to be supplied by pipeline from Norman Wells. This spectacular project took twenty months, 25 000 men, 11 million tonnes of equipment, 1 600 kilometres of road, 1 600 kilometres of telegraph line and 2 575 kilometres of pipeline. Dubbed Canol, the name was probably a contraction of the longer title "Canadian Oil." Estimates of the project's cost range up to $300 million.[16] The pipeline network consisted of the 950-kilometre crude oil line from Norman Wells to Whitehorse and three lines to carry products to Skagway and Fairbanks, Alaska, and Watson Lake, Yukon.

Because of wartime urgency, the line ran on top of the ground, alongside the road, often without supports. Vulnerable to frost heaving, snowstorms and flooding, the Canol pipeline benefitted from few of the normal precautions commonly employed today by pipeline construction companies. The pipe was not designed for extreme cold and, without proper handling and installation, failed frequently.[17]

Meanwhile, Imperial drilled more wells. The test for the Norman Wells oil field came when the pipeline began operating on February 16, 1944. The field surpassed expectations. During the remaining year of the Pacific war, the pipeline pumped about 160 000 cubic metres of oil to the Whitehorse refinery. But at the end of the war the line had no commercial value. Imperial bought the Whitehorse refinery for $1 million, dismantled it and moved it to Edmonton. There, the company reassembled the facility to handle production from the post-war Alberta discoveries.

OTHER SUCCESSES AND DISAPPOINTMENTS

The goal of most petroleum exploration in this era was oil. As the automobile grew in popularity worldwide, Canada emerged as one of its most enthusiastic converts. Considering the country's size, this was no surprise. The climate also dictated heavy use of fuel oils for heating. This demand, filled in large part from outside the country, put pressure on oil explorers to succeed. And they did find oil. Generally, they found it in lower Cretaceous reservoirs – small quantities of heavy oil that were either uneconomical to produce or quickly exhausted.[18]

In the course of this search, the explorers also found gas. They shut in most of these gas discoveries but occasionally a field developed close to a town or city that provided a local market. A small amount of exploration sought sources of natural gas in order to support regional natural gas distribution systems in Alberta.

In a few cases, town councils coveted cheap sources of local fuel. Some fell prey to promoters and drillers eager for contracts and not too worried about the presence or absence of promising geological structures. But there were important natural gas developments in this era. The discovery of the Viking-Kinsella field justified a pipeline to Edmonton in 1923. A gas find at Lloydminster, in 1934, provided fuel for that border community. The 1944 discovery of wet sour gas at Jumping Pound eventually supplied part of Calgary's needs and tied Exshaw and Banff into the natural gas distribution system. The Jumping Pound field tapped the prolific Mississippian limestone. In developing it, Shell drilled what was then the deepest well in Canada. Its Jumping Pound #4 well reached a depth of more than 4 000 metres.

Pre-World War II oil finds outside Turner Valley included the discovery of heavy oil at Wainwright in 1925 and at Lloydminster in 1939. These led to the construction of local refineries.

The leader in both exploration activity and frustration in this era was Imperial Oil. Looking for another Turner Valley, the company drilled a series of wells, finding a little gas here, a little heavy oil there and a lot of nothing everywhere else. The big one eluded Imperial before, during and immediately after World War II. By 1946, it seemed possible that the portion of the vast Western Canadian Sedimentary Basin straddling the four western provinces contained no major oil fields other than the unusual discovery at Turner Valley.

Chapter 5

The Leduc Era

Rigid Drilling Co. employees and relatives near Leduc, 1947.
(Provincial Archives of Alberta, H. Pollard Collection, P1400)

It takes skill to drill wells in Alberta and not hit oil or gas. Skill, or bad luck. Fortune was not with Imperial Oil as it searched the prairies after World War II. After 133 dry holes in Saskatchewan and Alberta, Imperial almost gave up its drilling program and moved elsewhere for the next major oil field in Canada. But in 1946, the company embarked on one last wildcat drilling program across Alberta. It began with Leduc #1 on Mike Turta's farm, 15 kilometres west of Leduc and 50 kilometres south of Edmonton. Located on a weak seismic anomaly and 80 kilometres from the closest attempt to find oil, it was a "rank wildcat."

LEDUC #1 AND #2

Drilling at Leduc #1 started November 20, 1946, and continued through a winter described as "bloody cold" by members of the crew. It looked like a gas well at first but, at about 1 530 metres, drilling sped up and bit samples began to show hints of oil. At 1 544 metres, oil flowed to the surface. Imperial finally had an oil discovery and the company decided to bring the well in with fanfare. It invited the mayor of Edmonton and other dignitaries to a celebration at 10 o'clock on the morning of February 13, 1947. The night before the big event, swabbing equipment broke down, forcing the crew to labor through the night to repair it.

The appointed hour came and no oil flowed. Many of the invited guests went home. Finally, at 4 p.m., the mud blew out of the hole and the chilled onlookers, by then numbering about 100, watched a spectacular column of smoke and fire as the first gas and oil flared in the evening sky. Alberta's Minister of Mines, Nathan Tanner, opened a valve and the Canadian oil industry burst into the modern era.

Imperial lost no time developing its find. Almost immediately the company began drilling Leduc #2, about three kilometres southwest of #1, hoping to delineate the producing formation. Nothing showed at the expected depth, however, and company officials disagreed over what to do next. One group wanted to abandon the well and drill a direct offset to #1. Another group voted to continue drilling #2 as a deep stratigraphic test. Drilling continued while the arguments raged. Then, on May 10, Leduc #2 struck the vastly bigger Devonian reef formation at 1 657 metres, 100 metres deeper than the discovery at #1. This formation became one of the most prolific in Canada.

EXPLORING THE DEVONIAN REEFS

The Leduc discovery put Alberta on the world petroleum map. News of the field spread quickly, due in large part to a spectacular blowout in the early days of the development of this field. In March 1948, drillers on the Atlantic Leduc #3 well lost mud circulation in the top of the reef, and the well blew out. In one journalist's words,

> The well had barely punched into the main producing reservoir a mile below the surface when a mighty surge of pressure shot the drilling mud up through the pipe and 150 feet into the air. As the ground shook and a high-pitched roar issued from the well, the mud was followed by a great, dirty plume of oil and gas that splattered the snow-covered ground. Drillers pumped several tons of drilling mud down the hole, and after 38 hours the wild flow was sealed off, but not for long. Some 2 800 feet below the surface, the drill pipe had broken off, and through this break the

pressure of the reservoir forced oil and gas into shallower formations. As the pressure built up, the oil and gas were forced to the surface through crevices and cracks. Geysers of mud, oil, and gas spouted out of the ground in hundreds of craters over a ten-acre area around the well.[1]

Atlantic #3 eventually caught fire, and the crew worked frantically for 59 hours to snuff out the blaze.

It took six months, two relief wells and the injection of 700 000 barrels of river water to bring the well under control. Cleanup efforts recovered almost 1.4 million barrels of oils in a series of ditches and gathering pools. The size of the blowout and the cleanup operation added to the legend. By the time Atlantic #3 was back under control, the whole world knew from newsreels and photo features of the blowout that the words "oil" and "Alberta" were inseparable.

Exploration boomed. By 1950, Alberta was one of the world's exploration hot spots, and seismic activity grew until 1953. After the Leduc strike, it became clear that Devonian reefs could be prolific oil reservoirs, and exploration concentrated on the search for similar structures. A series of major discoveries followed, and the industry began to appreciate the diversity of geological structures in the province that could contain oil. Early reef discoveries included Redwater in 1948, Golden Spike in 1949, Wizard Lake, Fenn Big Valley and Bonnie Glen in 1951 and Westerose in 1952. In 1953, drillers found Pembina, the largest field in western Canada, in a sandstone formation. By 1956, more than 1 500 development wells dotted the Pembina field, with hardly a dry hole among them. The Swan Hills field, discovered in 1957, exploited a carbonate rock formation.[2]

Geologists only looked for oil. The price of gas remained low, with limited markets. But major gas discoveries appeared at Pincher Creek in 1948, Cessford in 1950, Bindloss, Hussar, Minnehik, Duck Lake, Nevis and Olds in 1952.

EXPLORATION OUTSIDE ALBERTA

The widening scope of exploration in Alberta soon led to investigation of basin lands outside the province. To the east, exploration spilled over into the southern half of Saskatchewan and into southwestern Manitoba where, in 1951, an important light oil discovery greeted drilling efforts near Virden. The Virden discovery sparked rapid development in nearby areas of both Manitoba and Saskatchewan. Numerous heavy oil discoveries followed in the Lloydminster area of Saskatchewan, with light oil finds in other parts of the province. Exploratory drilling in Saskatchewan peaked for the decade in 1954.

In 1919, British Columbia withdrew its lands from exploration by private interests. Although lifted briefly in 1933, this ban lasted for 25 years in the Fort St. John area, longer in other parts of the province.

In 1921-22, the provincial government conducted a six-well program in the Peace River area of northeastern B.C. and found small accumulations of natural gas and oil, all non-commercial. And in 1924, an Imperial Oil subsidiary made an important gas discovery just across the Alberta border near Pouce Coupe. Although there were no follow-ups to these discoveries at the time, drilling in the 1940s led to confirmed substantial gas reserves for the area and eventually led to the development of the Westcoast Transmission natural gas pipeline. Subsequent exploration in B.C. yielded unusually high success rates for wildcat wells. But until the Boundary Lake oil discovery in 1955, only gas was found in significant volumes.

In the older exploration areas of central and eastern Canada, activity met limited success in southern Ontario along the St. Lawrence lowlands and in New Brunswick. Although discoveries proved modest, many small finds had economic value. It took until 1959 for oil production in Ontario to exceed the record set in 1895. From that point on, Ontario set new records every year until 1967, when oil production peaked at around 1.4 million barrels. Ontario natural gas production, however, had already peaked in 1917.

In the St. Lawrence lowlands of Quebec, small natural gas reservoirs supplied fuel for local consumption. In 1960, on land owned by a religious order, *Les pères de la Fraternité Sacerdotale,* a crew drilled a shallow well at Pointe du Lac near Trois-Rivières. The order hoped for a local source of inexpensive fuel, primarily to heat its monastery. The well encountered high-pressure natural gas at only 80 metres and blew wild. The flow rate could not be measured but was estimated at between 17 and 35 million cubic feet per day – remarkable volumes even by Alberta standards. The gas flowed unfettered during attempts to control the flow, but it blew wild for nearly two months. Finally, a relief well near the discovery well plugged it with concrete. The incident was an anomaly. The Pointe du Lac gas field's remaining reserves were only about three billion cubic feet, but the field remained a small producer until its reserves depleted in the mid-1970s.

MOVING THE PRODUCT TO MARKET

The discoveries of oil and gas in the Leduc era were so large that local markets could not begin to absorb the supply. Producers needed pipelines to take western Canadian petroleum to the heavily-populated central region of Canada and or the United States. Oil pipelines materialized quickly while those for natural gas took time. Concerns arose regionally and nationally about future gas supplies, and development stalled for years. The feverish exploration spawned by Leduc coincided with a period of export hearings and heated political debate over who had the right to sell, transport and buy western Canadian gas.

Chapter 6

The Pipeline Era

Shipping crude oil for the Wainwright oilfield by tank car, no date. Note pump in front of horses, which is being used to transfer the crude from truck's tank to tank car. (Glenbow Archives/NA-544-132)

In 1853, a small gas transmission line in Quebec established Canada as a leader in pipeline construction. A 25-kilometre length of cast-iron pipe moved natural gas to Trois-Rivières, to light the streets. It was probably the longest pipeline in the world at the time.[1] Canada also boasted the world's first oil pipeline when, in 1862, a line connected the Petrolia oil field to Sarnia, Ontario.[2] In 1895, natural gas began flowing to the United States from Ontario's Essex field through a 20-centimetre pipeline laid under the Detroit River.

In western Canada, Eugene Coste built the first important pipeline in 1912. The 274-kilometre natural gas line connected the Bow Island gas field to consumers in Calgary. Canada's debut in northern pipeline building came during World War II when the short-lived Canol line delivered oil from Norman Wells to Whitehorse (964 kilometres), with additional supply lines to Fairbanks and Skagway, Alaska, and to Watson Lake, Yukon. Wartime priorities assured the expensive pipeline's completion in 1944 and its abandonment in 1946.

By 1947, only three Canadian oil pipelines moved product to market. One transported oil from Turner Valley to Calgary. A second moved imported crude from coastal Maine to Montreal while the third brought American mid-continent oil into Ontario. But the Leduc strike and subsequent discoveries in Alberta created an opportunity for pipeline building on a grander scale. As reserves increased, producers clamored for markets. With its population density and an extensive refining system that relied on the United States and the Caribbean for crude oil, Ontario was an excellent prospect. The west coast offered another logical choice – closer still, although separated from the oil fields by the daunting Rocky Mountains. The industry pursued these opportunities vigorously.

THE CRUDE OIL ARTERIES

Construction of the Interprovincial Pipeline Ltd. system from Alberta to Central Canada began in 1949 with surveys and procurement. Field construction of the Edmonton/Regina/Superior (Wisconsin) leg began early in 1950 and concluded just 150 days later. The line began moving oil from Edmonton to the Great Lakes, a distance of 1 800 kilometres, before the end of the year. In 1953, the company extended the system to Sarnia, Ontario, and in 1957 to Toronto. Until the completion of the TransCanada gas pipeline, Interprovincial (IPL) was the longest pipeline in the world.

The IPL line fundamentally changed the pricing of Alberta oil to make it sensitive to international rather than regional factors. The wellhead price reflected the price of oil at Sarnia, less pipeline tolls for shipping it there. IPL is by far the longest crude oil pipeline in the western hemisphere.[3] Looping, or constructing additional lines beside the original, expanded the Interprovincial system and allowed its extension into the American Midwest and to upstate New York. In 1976, it reached its present length of 3 680 kilometres through an extension to Montreal. Although it helped assure security of supply in the 1970s, the extension became a threat to Canadian oil producers after deregulation in 1985. With Montreal refineries using cheaper imported oil, there was concern within the industry that a proposal to use the line to bring foreign oil into Sarnia might undermine traditional markets for Western Canadian petroleum.

The oil supply situation on the North American continent grew critical during the Korean War and helped promote construction of the Trans Mountain pipeline from Edmonton to Vancouver and, later, to the Seattle area. Oil first moved through the 1 200-kilometre, $93 million system in 1953.[4] The rugged terrain made the Trans Mountain line a feat of extraordinary engineering. It crossed the Rockies, the mountains of central British Columbia, and 98 streams and rivers. Where it crosses under the Fraser River into Vancouver at Port Mann, 700 metres of pipe lie buried nearly five metres below the river bed.[5] At its highest point, the pipeline is 1 200 metres above sea level.

To support these major pipelines, the industry gradually developed a complex network of feeder lines in the three most westerly provinces. A historic addition to this system was the 866-kilometre Norman Wells pipeline. This pipeline accompanied the expansion and waterflooding of the oil field, and began bringing 2 500 barrels of oil per day to Zama, in northwestern Alberta, in early 1985. From Zama, Norman Wells oil travels through other crude oil arteries to Alberta and other Canadian refineries.

THE POLITICS OF NATURAL GAS TRANSPORTATION

Those who applied for permits to export Alberta natural gas made the painful discovery that it was politically more complex to export gas than oil. Canadians tend to view oil as a commodity. However, through much of Canadian history, they have viewed natural gas as a patrimony, an essential resource to husband with great care for tomorrow. Although the reasons behind this attitude are complex, they are rooted in an incident at the turn of the century, when Ontario revoked export licences for natural gas to the United States.

By the late 1940s Alberta, through its Conservation Board, eliminated most of the wasteful production practices associated with the Turner Valley oil and gas field. As new natural gas discoveries greeted drillers in the Leduc-fueled search for oil, the industry agitated for licences to export natural gas. In response, the provincial government appointed the Dinning Natural Gas Commission to inquire into Alberta's likely reserves and future demand.

In its March 1949 report, the Dinning Commission supported the principle that Albertans should have first call on provincial natural gas supplies, and that Canadians should have priority over foreign users if an exportable surplus developed. Alberta accepted the recommendations of the Dinning Commission, and later declared it would only authorize exports of gas in excess of a 30-year supply. Shortly thereafter, Alberta's Legislature passed the Gas Resources Conservation Act, which gave Alberta greater control over natural gas at the wellhead, and empowered the Oil and Gas Conservation Board to issue export permits.[6]

The federal government's policy objectives at the time reflected concern for national integration and equity among Canadians. In 1949, Ottawa created a framework for regulating interprovincial and international pipelines with its Oil or Gas pipeline act.[7] Alberta once again agreed to authorize exports. The federal government, like Alberta, treated natural gas as a Canadian resource to protect for the foreseeable future before permitting international sales.

Although Americans were interested in Canadian exports, they wanted only cheap natural gas. After all, their natural gas industry was a major player in the American economy and American policy-makers were not eager to allow foreign competition unless there was clear economic benefit. Consequently, major gas transportation projects were politically and economically uncertain.

Wagon carrying pipe for the Bow Island to Calgary natural gas pipeline construction project, 1913.
(Glenbow Archives/NA-4048-1)

GAS PIPELINE CONSTRUCTION BEGINS

Construction of Turner Valley to Calgary pipeline, 1920s.
(Provincial Archives of Alberta/PAA-85.248 Box 39)

Among the first group of applicants hoping to remove natural gas from Alberta was Westcoast Transmission Limited, backed by British Columbia-born entrepreneur Frank McMahon. The Westcoast plan, eventually achieved in a slightly modified form, took gas from northwestern Alberta and northeastern B.C. and piped it to Vancouver and the American Pacific northwest, supplying B.C.'s interior along the way. Except for a small export of gas to Montana which began in 1951, Westcoast was the first applicant to receive permission to remove gas from Alberta. Although turned down in 1951, Westcoast received permission in 1952 to take 50 billion cubic feet of gas out of the Peace River area of Alberta annually for five years. The company subsequently made gas discoveries across the border in B.C. which further supported the scheme. However, the United States Federal Power Commission (FPC), today known as the Federal Energy Regulatory Commission, rejected the Westcoast proposal in 1954 after three years of hearings and 28 000 pages of testimony.

Within 18 months, however, Westcoast returned with a revised proposal, found a new participant in the venture, and received FPC approval. Construction began on Canada's first major gas export pipeline.

The Canadian section of the line cost $198 million to build and at the time was the largest private financial undertaking in the country's history. Built in the summer seasons of 1956 and 1957, the line moved gas from the Fort St. John and Peace River areas 1 250 kilometres to Vancouver and the American border.

Two applicants originally expressed interest in moving gas east from Alberta. Canadian Delhi Oil Company proposed moving gas to the major cities of eastern Canada by an all-Canadian route, while Western Pipelines wanted to stop at Winnipeg with a branch line south to sell into the midwestern United States. In 1954 C.D. Howe forced the two companies into a shotgun marriage, as TransCanada PipeLines, with the all-Canadian route preferred over its more economical but American-routed competitor.

This imposed solution reflected problems encountered with the construction of the Interprovincial oil pipeline. Despite the speed of its construction, the earlier line caused angry debate in Parliament, with the Opposition arguing that Canadian centres deserved consideration before American customers and that "the main pipeline carrying Canadian oil should be laid in Canadian soil".[8] By constructing its natural gas mainline along an entirely Canadian route, TransCanada accommodated nationalist sentiments, solving a political problem for the federal government.

The regulatory process for TransCanada proved long and arduous. After rejecting proposals twice, Alberta finally granted permission to export gas from the province in 1953. At first, the province waited for explorers to prove gas reserves sufficient for its 30-year needs, intending to only allow exports in excess of those needs. After clearing this hurdle, the federal government virtually compelled TransCanada into a merger with Western Pipelines. When this reorganized TransCanada went before the Federal Power Commission for permission to sell gas into the United States, the Americans greeted it coolly: the FPC was sceptical of the project's financing and unimpressed with Alberta's reserves.

Engineering problems made the 1 090-kilometre section crossing the Canadian Shield the most difficult leg of the TransCanada pipeline. Believing construction costs could make the line uneconomic, private sector sponsors refused to finance this portion of the line. Since the federal government wanted the line laid for nationalistic reasons, the reigning Liberals put a bill before Parliament to create a crown corporation to build and own the Canadian Shield portion of the line, leasing it back to TransCanada. The government restricted debate on the bill in order to get construction underway by June, knowing that delays beyond that month would postpone the entire project a year. The use of closure created a furore which spilled out of Parliament and into the press. Known as the Great Pipeline Debate, this parliamentary episode contributed to the government's defeat at the polls in 1957.

But the bill passed and construction of the TransCanada pipeline began. The completion of this project was a spectacular technological achievement. In the first three years of construction (1956-58), workers installed 3 500 kilometres of pipe, stretching from the Alberta-Saskatchewan border to Toronto and Montreal. Gas service to Regina and Winnipeg commenced in 1957 and the line reached the Lakehead in Ontario before the end of that year. In late 1957, during a high pressure line test on the section of the line from Winnipeg to Port Arthur (today called Thunder Bay), about five and a half kilometres of pipeline blew up near Dryden, Ontario. After quick repairs, the line delivered Alberta gas to Port Arthur before the end of the year, making the entire trip on its own wellhead pressure.

Building the Canadian Shield leg required continual blasting. For one 320-metre stretch, the construction crew drilled 2.4 metre holes into the rock, three abreast, at 56-centimetre intervals. Dynamite broke up other stretches, 305 metres at a time.[9]

On October 10, 1958, a final weld completed the line and on October 27, the first Alberta gas entered Toronto. For more than two decades, the Trans-Canada pipeline was the longest in the world. Only in the early 1980s was its length finally exceeded by a Soviet pipeline from Siberia to western Europe.

TransCanada and Westcoast Transmission received their Alberta gas at the province's borders. The Alberta Gas Trunk Line (AGTL) system, now called NOVA Corporation of Alberta, gathered gas from wells in the province and delivered it to exit points.

There were many reasons for the creation of AGTL. One was that the provincial government considered it sensible to have a single gathering system in Alberta to feed export pipelines, rather than a number of separate networks for TransCanada, Westcoast and the other proposed export pipelines.

Another was that pipelines crossing provincial boundaries and those leaving the country fall under federal jurisdiction. By creating a separate entity to carry gas within Alberta, the provincial government stopped Ottawa's authority at the border.

Incorporated in 1954, AGTL issued public shares in 1957. The 2.5 million shares, reserved to Alberta buyers, met a high demand. Despite a limit of 100 shares per customer, Albertans oversubscribed by 500 percent. Shares quickly doubled in value.

Legislation originally limited AGTL's mandate to gathering gas within the province. In 1974, with a broadened role, AGTL quickly became a highly diversified company. Today the company operates more than 18 000 kilometres of pipeline, is the majority owner of a Calgary-based integrated oil company, is one of the world's largest producers of petrochemicals and has business interests in manufacturing and high technology. To reflect its newfound diversity, AGTL changed its name to NOVA Corporation in 1980.

GAS EXPORT PIPELINES

The five pipelines described so far – Interprovincial, Trans Mountain, Westcoast, TransCanada and the Nova system – provided Canada with a basic transportation infrastructure for its petroleum industry. After their completion, the next major pipeline development efforts were made by TransCanada and a new company, Alberta and Southern Gas Company (A&S), during the fifties. Both companies wished to gain permission to market gas into the United States but they wanted to serve different regional markets. TransCanada wanted to serve the American Midwest through the Emerson, Manitoba, border gate while A&S sought to market gas from Alberta's foothills trend reservoirs, conveying it to an American pipeline. The stateside pipeline would be owned by the A&S parent company, Pacific Gas and Electric of San Francisco.

Alberta granted both companies permission to proceed with these export schemes in 1959, and the newly-created National Energy Board approved their proposed exports in early 1960. To serve the A&S system, AGTL built a trunk line down the foothills trend from a series of processing plants spurred into existence by the export approval. From the Alberta border, the gas passed through a pipeline in British Columbia to a Kingsgate, Idaho, export point, and on to California. At the beginning of 1962, Alberta gas began moving to California through the world's largest-diameter pipeline.

Thirty years later, however, the natural gas industry was in a very different position. As concern over oil related pollution rose in the 1980s, natural gas became increasingly attractive to producers and consumers alike. When deregulation in the middle part of the decade flooded the markets with gas, analysts predicted a short period of excess supply followed by shortages. But when the United States began offering attractive subsidies to producers of coal bed gas, the supply glut increased. Consumer agencies traded cost consciousness for their previous concern over security of supply and by the early 1990s new pipeline projects proliferated to help move natural gas to distant North American markets.

In the late 1980s, plans were drawn up for a 534-kilometre pipeline to move gas from the British Columbia mainland to Vancouver Island. Although subsidized by various levels of government, politicians lauded it as a way to reduce the island's dependence on coal, oil and hydro-electric power. Delays in the projected 18 month construction schedule and cost overruns were partly due to environmental concerns but a federal review approved the project. By late 1991, supplies gas began flowing to consumers on the island.

In 1990, the Iroquois Gas Transmission System received permission from the American Federal Energy Regulatory Commission and the NEB to build a $583 million pipeline. Using western Canadian gas from the TransCanada pipeline system, it began supplying markets in the American northeast in 1992 after crossing the St. Lawrence River into New York state at Iroquois, Ontario.

During 1993, various Canadian and American companies collaborated on the Pacific Gas expansion project. Using additional pipeline volume and compression facilities, they added 75 percent more capacity to the Alberta-to-California pipeline system, which already carried 1.2 billion cubic feet of gas per day. The system also carried an additional 150 million cubic feet per day to the Pacific northwest and 750 million cubic feet more per day to California. The project also extended the pipeline from northern California south to the Los Angeles area.

As long as gas surpluses exist in Canada and customers further south find the prices attractive, proposals for new gas pipelines will appear regularly in the business news. In 1992 Altamont, a Houston-based company, failed in its attempt to build a competing natural gas pipeline to California. After the National Energy Board rejected its application, Altamont deferred its project until a more propitious time.

PIPELINES FROM THE FRONTIERS

The prolific Prudhoe Bay oil strike in Alaska in 1968 affected the Canadian oil industry in many ways, a number of which are discussed elsewhere in this book. As far as the pipeline business is concerned, the Prudhoe Bay oil strike led indirectly to two Canadian projects competing to carry Prudhoe Bay gas through Canada to the United States.

The first of these projects originated as a plan to transport Alaskan oil by pipeline through Canada. A consortium of companies suggested this proposal as an alternative to the proposed Alyeska pipeline which eventually transported oil along an all-American route, first by pipeline to Alaska's Pacific coast at Valdez and then by tanker to markets in the lower 48 states.

The companies behind the Canadian proposal formed the Mackenzie Valley Pipeline Research Group in 1969 to study the feasibility of an oil pipeline through eastern Alaska to northern Alberta. Both AGTL (Nova) and TransCanada were important members of this group. In the summer of 1970, the group turned its attention to moving Prudhoe Bay gas through the Yukon to the Mackenzie River delta, up that river valley to Alberta, and on to the United States.

In June 1970, AGTL surprised the group by announcing its withdrawal from the consortium and proposing a competing plan to move Mackenzie delta to Alberta. Originally, the AGTL proposal was for an all-Canadian project, and would not move American natural gas at all. A major selling point from the industry's point of view was that this line would be a common carrier, available to any shipper on a first-come, first-served basis. The Canadian portion would be owned exclusively by Canadian corporate and individual shareholders. The Canadian ownership position made it known as the "Maple Leaf Pipeline." The company later concluded that supplies and markets would not justify the pipeline, and began a head-to-head competition with the Arctic Gas Pipeline group by proposing a pipeline that would roughly parallel the Alaska Highway, crossing southern Alaska and the southern Yukon to northern British Columbia. In northern B.C. some of the gas would be transported through existing Westcoast Transmission facilities for delivery to westcoast markets.

The pipeline would continue eastward from this juncture to the Alberta border, where it would connect with existing AGTL facilities in northern Alberta. From that border point it would flow to American midwestern markets through a new pipeline from southern Alberta to the American Midwest. This proposal – which initially did not make any provision for shipping Mackenzie delta gas – was called the Alaska Highway Pipeline.

As AGTL developed its proposals, the Arctic Gas Pipeline group completed a definitive project of its own, and a long contest ensued before regulatory authorities. The competing proposals received scrutiny from the National Energy Board, the United States Federal Power Commission and, eventually, from a Canadian federal commission of inquiry known as the Berger Commission. As hearings and inquiries progressed, both contestants amended their applications to provide for potential exploitation of gas discoveries in the delta of the Mackenzie River, as Canadian explorers found important new gas fields in the exploration frontier. Both groups made proposals to "piggyback" Mackenzie Delta gas along with Alaska gas, through lateral connection to the main pipeline.

As the competition heated up, several organizations raised concerns about the impact of such a major development on the northern environment and on the lifestyle of the native people. The federal government's response came in 1974 when it appointed the Mackenzie Valley Pipeline Inquiry, a commission headed by British Columbian Justice Thomas Berger. Much to the surprise of the industry, Berger became extremely involved in the issues surrounding the pipeline – such issues as foreign ownership and nationalism, native land claims and the environmental fragility of the north.

In 1977, after extensive hearings in northern communities, the Berger Commission recommended a 10-year postponement of Mackenzie Valley pipelines and a permanent ban on pipeline development along the north slope of the Yukon. As an alternative, the Commission recommended approval of a new proposal from the sponsors of the Maple Leaf Pipeline to construct a multi-billion dollar pipeline along the existing route of the Alaska Highway, a pipeline which would carry only Alaska gas with delta gas connected later. The National Energy Board later endorsed this recommendation and the government authorized it by order-in-council.

The governments of Canada and the United States later signed an international agreement to develop the Mackenzie Valley pipeline. To provide early cash flow for the project, two southern legs of the pipeline, the "pre-build pipelines," received approval in July 1980 for immediate construction. Upon completion in 1982, the pre-build sections could carry up to 750 million cubic feet of natural gas per day from Alberta to markets in the United States. Designed for large volumes of Alaska gas, these sections have never worked to full capacity. Major new natural gas discoveries in Alberta and British Columbia, combined with the recession of the early 1980s and soft markets for natural gas in the United States, brought development of the northern portion of the Alaska Gas Pipeline to a halt. Completion of this project awaits an increase in the price of natural gas.

Other proposals of the 1970s and the early 1980s to carry frontier gas to market also fell victim to poor market conditions. One, the Polar Gas Pipeline, would have delivered natural gas from the Arctic Islands to southern markets. Polar Gas amended its proposal many times. In late 1984, the consortium filed an application with the National Energy Board for an express 920-millimetre gas pipeline from the Mackenzie Delta to Edson, Alberta, where it would connect with the Nova system.

Two proposals were made to develop liquefaction facilities in the Arctic Islands and transport liquefied natural gas to North American and European markets. These also died in the early eighties when surpluses of both oil and gas on international markets raised questions about their commercial viability. A third liquefaction project proposed transporting natural gas to Japan from British Columbia, and received conditional approval from the National Energy Board in 1982. World energy surpluses also postponed that project in 1986.

Companies have also proposed pipelines to carry gas from east coast offshore discoveries. One proposal advocates a pipeline and production facilities to transport offshore natural gas from the Sable Island area off Nova Scotia. Although stalled by high costs, that project could have transported 300 million cubic feet of natural gas each day from fields 210 kilometres offshore through a subsea, concrete-coated pipeline. On land, a refinery would have stripped the gas of natural gas liquids and processed it for delivery to commercial pipelines. Then a multi-billion dollar pipeline would have transported the gas to Maritime and northeastern American markets.

OTHER KINDS OF PIPELINES

In addition to the large pipelines, there are gathering lines within the fields, gas distribution lines within the urban areas, and lines carrying refinery products to points from which distribution continues by rail and truck. Also, there are pipelines, some as long as major oil and gas lines, which carry natural gas liquids, other more volatile gas processing products and liquid petrochemicals.

Among the earliest Canadian products pipelines were the three built by Imperial in 1954 to carry propane, butane and pentanes plus from the Leduc gas conservation plant to Edmonton. The first long-distance products pipeline in Canada was a 933-kilometre line from the Empress straddle plant on the Alberta/Saskatchewan border to Winnipeg, constructed in the early 1960s. Another was the Cochrane to Edmonton (Co-Ed) pipeline. Designed to meander through the Alberta countryside, the pipeline collects natural gas liquids from processing plants. Conveyed to Edmonton, these liquids are "batched" through the Interprovincial pipeline to Sarnia.

The Cochin pipeline was another major development of the late 1970s. Approximately 3 100 kilometres in length, it carries petrochemical and high vapor pressure processing products from Fort Saskatchewan, Alberta, to Sarnia, Ontario and on to Green Springs, Ohio. Supplied with product through the 880-kilometre Alberta ethane gathering system, the Cochin line terminates at a large fractionation and distribution complex in Sarnia.

By the 1990s, the era of big pipelines in Canada was four decades old. With oil and gas supplies typically far removed from markets, pipelines became a phenomenon of the second half of the century, allowing large amounts of oil and gas to reach distant markets for the first time. After moving product to local and regional markets, continental arteries carried petroleum to nearly every urban community in North America.

Oil lines came first, then gas. Always more politically sensitive, natural gas exports for decades required large stocks of reserves for Canadian consumers before governments licensed exports. Eventually, however, even gas pipelines became continental in scope, moving the valuable commodity from the cold reaches of the north to the warmer southern provinces and states. By the 1990s, price decontrol created surpluses of both oil and gas, resulting in demands for extremely competitive pricing as consumers forgot the shortages of the mid-1970s.

Pipeline construction was critical to the development of the petroleum industry. But almost equally critical for the natural gas segment of the industry was natural gas processing. After all, most natural gas requires processing before entering the pipelines.

Chapter 7

Processing Gas

Shell Canada's Caroline gas plant during construction, 1992.
(Shell Canada)

For decades, natural gas was a nuisance. Dangerous to handle and hard to get to market, early oilmen despised it as a poor relation to its rich cousin oil. But processing changes the commodity in two critical ways. First, it extracts valuable by-products; second, it renders natural gas fit for transportation and commercial sale. Because of ever-evolving technology, the modern gas processing industry extracts higher percentages of a wider range of hydrocarbons and by-products than its predecessors and removes higher percentages of dangerous and other unwanted impurities.

CANADA'S FIRST LIQUID EXTRACTION PLANT

In the nineteenth century, commercial demand for natural gas was so limited that discoveries were often only developed if consumers could use the gas just as it came out of the ground. Early processing removed water. If the gas required further processing, the producer shut in the well. Flares got rid of gas coming from an oil well.

Natural gas processing began in earnest with the advent of the internal combustion engine. The condensate or naphtha found in "wet" gas reservoirs served as engine fuel. The first gas processors employed cooling and compression to obtain the liquids. They were volatile when they also included petroleum gases (LPGs), propane and butane, and some of the heavier hydrocarbons. Farmers often used these natural gas liquids in their early tractors. While some car operators chose to use this "wild stuff" directly as automobile fuel in the Model T and Model A Fords of the day, it made for a jerky and undependable ride.

The earliest gas processing occurred in the United States. In 1913, Hope Natural Gas built the first oil absorption plant for gas liquids extraction at Hastings, West Virginia. The technology of the day sprayed the gas stream with oil to absorb liquids present in the gas. Distillation stripped the resulting "rich" oil stream of its gas liquids. One year after the Hastings plant went into operation, a similar processing plant made its Canadian debut at Turner Valley. The entrepreneurs behind the 1914 Dingman wet gas discovery built an expensive oil absorption plant to harvest the liquids from the gas, but the plant burned down in 1920. Royalite, as the new owner, replaced the plant in 1921.

Removing marketable by-products from raw natural gas is one part of gas processing. The other part involves stripping impurities from the gas to allow for its safe transportation and consumption.

CANADA'S FIRST SWEETENING PLANT

Much gas is "sour" or laced with dangerous hydrogen sulphide (H_2S) in its natural state. The process of taking out this impurity is called "sweetening." At low concentrations hydrogen sulphide has an obnoxious rotten egg smell. This odor probably annoyed Union Gas Company of Toronto customers and prompted the company to build Canada's first sweetening plant in 1924 at Port Alma, Ontario, to "scrub" Tillbury gas. It removed hydrogen sulphide by exposing the sour gas to dissolved soda ash. Although previously used on coal gas, the application at the Port Alma plant was the first time this process sweetened natural gas.

The second Canadian sweetening plant followed a year later in Turner Valley, and used the same process. The first gas at Turner Valley was sweet but the Royalite #4 discovery of 1924, from a deeper horizon, was sour.

PROCESSING GAS

Royalite built the Turner Valley sweetening plant in order to sell its gas to Canadian Western Natural Gas for distribution. But the technology of the day did not render the hydrogen sulphide harmless. Instead, the system simply disposed of the substance by diluting and dispersing it into the air from two tall stacks. Since H_2S is heavier than air, it settled back down, dispersed enough to be less than lethal, but smelling less than rosy. Turner Valley had a rotten egg bouquet on most days.

Burning gas at Hell's Half Acre, Turner Valley, 1926.
(Glenbow Archives/NA-1716-5)

IMPROVED HYDROCARBON EXTRACTION IN TURNER VALLEY

Between 1924 and 1927, Royalite operated two gas processing facilities side by side in Turner Valley: the sweetening plant and the liquids extraction plant.

The liquids extraction plant closed in 1927 and reopened in 1933 after the company revamped the facility. The new plant used "lean oil" absorption, a process that forced raw gas into contact with lean oil in chains of steel bubble caps. Improvement of the absorption medium and contact between the gas and the oil made for substantially higher rates of liquids recovery. The new plant was so successful that other companies built two similar plants in Turner Valley, and Royalite built a second plant to handle its production from the south end of the field. Gas and Oil Products Ltd. built a similar plant at Hartell in 1934 and British American (BA) opened one at Longview in 1936.

Once Alberta's Petroleum and Natural Gas Conservation Board began operating in 1938, the BA and Gas and Oil Products Limited plants had to change their operations significantly. Only Royalite had a market for its residue of gas stripped of liquids in the Canadian Western Natural Gas distribution system. The other two plants "flared" or burned off most of their residue gas until the board ruled that only wells connected to a market could be produced, thereby stopping the practice. Since the rule applied only to wells that tapped oil's overlying gas cap, the Hartell and Longview plants stayed in operation by processing solution gas, or gas produced along with oil from the Valley's wells.

GAS PROCESSING AFTER LEDUC: GAS CONSERVATION

As Alberta became an even larger oil producer after the Leduc discovery, the Conservation Board acted to prevent any repetition of the natural gas waste so common in Turner Valley. The board developed a broad conservation policy for natural gas. It prohibited producing natural gas from an oil reservoir's gas cap before the oil was fully produced, and included provisions aimed at conserving the natural gas often produced along with the oil. For this reason, these plants became known as "gas conservation plants."

The first of these new plants was Imperial's Leduc facility (sometimes called Imperial Devon or Imperial Leduc). It sweetened the gas with monoethanolamine (MEA), then extracted the liquid hydrocarbons by refrigeration. Northwestern Utilities Limited bought the gas at $14.12 per thousand cubic metres and distributed it in Edmonton. Trucks transported the propane, butane and pentanes plus extracted from the gas until 1954, when three pipelines began moving the products from Imperial Leduc to Edmonton. When markets couldn't be found for the propane, the board occasionally granted permission to flare it.

The next important plant built in Canada resulted from the discovery in 1944 of a wet sour gas find by Shell Oil at Jumping Pound, west of Calgary. Calgary, Exshaw (a small community where a cement plant became a large industrial consumer) and Banff were all potential markets for Jumping Pound gas, but the sour gas first required processing and sweetening. The gas plant began operating in 1951.

Built "California-style," with few buildings or other provisions for a cold climate, the original Jumping Pound plant ran into problems. During the first winter, water condensation and other cold weather problems led to one operational failure after another. When the second winter arrived, buildings sheltered most of the facilities. Shell Jumping Pound is sometimes referred to as Canada's "sour gas laboratory," for much of the industry's early understanding of sour gas processing came from experience there. It was the first sulphur plant in the world, its sulphur unit going into production in 1952. For this distinction it narrowly beat out the Madison Natural Gas plant which began extracting sulphur at Turner Valley later the same year.

GAS PROCESSING AFTER THE PIPELINES

As the Westcoast and TransCanada natural gas pipelines went into operation in 1957, a new and better day dawned for Canadian gas processing. Most of the gas that travelled those pipelines needed processing to meet the specifications of pipeline companies. Consequently, the late 1950s and early 1960s saw a boom in gas plant construction.

In 1957, a new gas plant at Taylor, near Fort St. John B.C., began supplying Westcoast Transmission. This plant differed from Alberta practices in a number of ways. For example, although it generally required dehydration, sweetening and processing for liquid hydrocarbons, companies transported the natural gas from northeastern British Columbia long distances before processing it further. Consequently, while planning the Westcoast pipeline, the field operators agreed to process all the gas at a single facility, rather than have individual gas plants in every major production area. At 10 million cubic metres per day, the Taylor plant had the capacity to process as much natural gas as all 11 of the other gas plants operating in Canada combined. The plant was also by far Canada's most northerly plant. Heavily insulated buildings protected the processing facilities and allowed them to function at temperatures typical of more southerly climes.

Handling sulphur at Madison natural gas facility, Turner Valley, 1952. (Provincial Archives of Alberta, H. Pollard Collection, P2973)

SELLING THE PRODUCTS

The enormous growth in Canadian processing capacity in the late 1950s and early 1960s brought great volumes of natural gas liquids and liquid petroleum gases to market. The natural gas liquids (NGLs) were seldom a problem because of their ready use in oil refining. Refiners also used butane for blending.

Propane presented a challenge because the volumes available greatly exceeded demand. Companies set out to widen the market with considerable success. Farmers and small communities not served by natural gas adopted it for home heating fuel.

In the early 1960s, markets for liquid petroleum gases grew rapidly. Companies responded by building "straddle" plants. These facilities straddled gas pipelines, to extract additional volumes of gas liquids from the gas stream. Where economic, field processors began "deep cutting" their own gas by installing facilities that culled more LPG from the gas through deep refrigeration. In the early 1970s, companies also began extracting the even lighter hydrocarbon ethane at some field processing and straddle plants. Ethane supplied Alberta's growing petrochemical industry as a feedstock for the manufacture of ethylene.

SULPHUR

From a slow start in 1952, sulphur production from gas processing snowballed as plant construction boomed in the late 1950s and early 1960s. Tough new regulations enacted by the Alberta government in 1960 forced the industry to reduce its emissions of such sulphur compounds as sulphur dioxide and hydrogen sulphide.

Over the years, sour gas processing technology steadily improved. By 1970, more stringent emission standards were technically feasible. The Alberta government announced new, tougher regulations in 1971. Improvements in sulphur extraction technology and the addition of tail gas clean-up units enabled processors to meet these stricter standards.

The amount of sulphur produced in Alberta increased rapidly, and soon far outstripped demand. By 1963, Alberta's annual sulphur production exceeded one million tons, compared with 30 000 tons in 1956. In 1973 it peaked at slightly more than 7 million tons.[1] Stockpiles grew annually. By 1978, 21 million tons of sulphur in large yellow blocks dotted the Alberta countryside. These inventories grew almost every year after 1952,[2] and government and industry became seriously concerned about the surplus. Since 1978, a strong sulphur marketing effort made Canada the largest supplier to international trade. Although sulphur sales were good during the 1980s, by the early 1990s a glut on the international market reduced its price.

PROCESSING GAS

Looking at the large, sophisticated, high-tech enterprise that Canadian gas processing is today, it is hard to imagine the challenges the industry faced as it developed. Until the major gas transmission pipelines began operating in the late 1950s, there was little incentive to develop a major gas processing sector.

SELLING TO A SURPLUS MARKET

During the 1980s and 1990s the natural gas industry faced a new series of problems. As demand for gas grew, suppliers expanded their capacity and soon a "gas bubble" developed. Although market analysts regularly forecast the end of the problem as only a few years away, the bubble refused to burst.

As a result, as crude oil prices dropped throughout the 1980s and natural gas reserves remained strong, consumers began taking advantage of the surplus. Forgetting the supply scares of the 1970s, individuals, corporations and governments alike shopped for the cheapest gas available.

Conservative governments in Washington D.C. and Ottawa alike moved their petroleum sectors towards deregulation in the mid-1980s. Throwing the market open to competition added to the surplus and depressed prices.

Suppliers across the continent began looking for new customers to make up in volume sales what they were unable to earn from low gas prices. But gas pipelines, built decades before, had little excess capacity.

Debate on a second gas pipeline from Alberta to California serves as a good example of the changing values during this period. For decades, California consumers opposed rival pipelines for fear of having to pay higher gas prices to cover pipeline construction. Deregulation made the pipeline companies common carriers so that any consumer group could buy space on the pipeline to move its gas. Gone were the days when the pipeline company moved the gas and also marketed it.

As deregulation put an end to vertically integrated gas delivery and marketing, consumers began crying for additional pipeline capacity. Half a continent away, cheap gas awaited. All they had to do was find a pipeline to move it. By the late 1980s and early 1990s, these groups and their elected officials supported throwing open the race to build pipelines to anyone. Competition among pipelines for the task of moving the gas to market promised to keep transmission costs reasonable.

As pipeline projects proliferated, natural gas producers sought new markets for their inexpensive and plentiful product. Electrical power generation with gas became a growth industry. As coal, hydroelectric and nuclear-powered generation facilities came under attack for environmental reasons, gas stepped in and sold itself as a clean alternative.

Entrepreneurs arranged for pipelines to transport natural gas, found markets for electricity, and even sold the heat created by the generator to another market. These "cogeneration plants" became popular. As long as the natural gas supplies exceed demands, these facilities remain attractive to producers and consumers alike. They use an inexpensive and environmentally friendly fuel. They meet immediate needs at only a fraction of the cost of large nuclear, hydro-electric or coal-powered facilities. Although their share of the market may shrink if gas prices rise, these ingenious projects fill an important market niche during a period of gas surplus.

The demand for vast supplies of natural gas to meet the expanding markets created a need for additional gas processing. For many decades, companies developed new gas fields in remote areas. But by the 1980s, the industry was reviewing its existing holdings, looking for the discoveries that eluded earlier exploration efforts.

The Caroline natural gas discovery in south-central Alberta in the mid-1980s brought the industry into a new era. As the biggest Canadian gas discovery since the 1970s and its richest gas project ever, the Shell-operated Caroline field stood out as a 10 billion dollar resource jewel. Although classified as a gas field, sulphur and other by-products from the gas promised to exceed the value of the natural gas itself.

But this rich discovery proved complicated, environmentally sensitive and economically challenging. The planning and review process took from 1986 to 1990, and set a new standard for community participation and consultation. Two companies, Shell and Husky, competed for the right to operate the field. In the process, farmers, acreage owners and other interested parties quickly made their concerns known. This development process forced corporations to compete for the right to develop the resource on new terms. Sustainable development theories came under close scrutiny, as did all aspects of the gas processing system.

Eventually, Shell and its backers won the bid, and developed a processing plant that recovers over 99 percent of the sulphur from the gas. But the Caroline experience made the public an integral part of the planning process and fine tuned the public consultation process. New risk assessment techniques also played a role in the process. In the end, the companies raised community relations to a new level as they recognized that the public consultation process was more important than ever.

By the early 1990s, natural gas processing had come of age. Since its infancy, when operators only removed a small part of the impurities, the gas sector has matured to become an important part of the petroleum industry. Gas moves around the North American continent in unprecedented volumes and is among the least expensive and most environmentally-desirable fuels. Cheap, abundant and popular, natural gas is the child of gas processing.

L. I. H. E.
THE BECK LIBRARY
WOOLTON RD., LIVERPOOL, L16 8ND

Chapter 8

Oil Sands
and the Heavy Oil Belt

Suncor bucketwheel excavating tar sand material, 1986. (David Finch)

It is difficult to grasp the immensity of Canada's oil sands and heavy oil resources.[1] The oil sands of northern Alberta include four major deposits which underlie almost 70 000 square kilometres of land – the Athabasca, Cold Lake, Peace River and Wabasca oil sands. The volume of bitumen in those sands dwarfs the light oil reserves of the entire Middle East. One deposit, the Athabasca oil sands, is the world's largest known crude oil resource.

Almost all Canadian bitumen is in Alberta. Most heavy oil is in Saskatchewan. What the oil industry calls the "heavy oil belt" is a string of reservoirs along the southern half of the Saskatchewan-Alberta border with the border town of Lloydminster as its buckle.

The range in density of petroleum is vast. The weight spectrum increases with the ratio of hydrogen to carbon in a compound's molecule. Methane (CH_4) has four hydrogen atoms for every carbon atom. That means it is light – a gas. Compounds with more carbon atoms for every hydrogen atom are heavier. Heavy oil and bitumen, which have more carbon than hydrogen, are heavy, black, sticky and either slow-pouring or so close to being solid that they will not pour at all unless heated. Although the dividing line is fuzzy, the term "heavy oil" refers to slow-pouring heavy hydrocarbon mixtures.

Suncor conveyor for transporting tar sand material to refinery, 1986.
(David Finch)

Suncor processing plant for extracting oil from tar sand material, 1986.
(David Finch)

Bitumen refers to mixtures with the consistency of cold molasses that pour at room temperatures with agonizing slowness.

Heavy oil is more fluid than bitumen, but its consistency varies. Conventional heavy oil is producible by standard drilling methods or by recovery methods which inject water into a well to create pressure and make the oil flow. Non-conventional heavy oil needs heat or solvents and pressure in the reservoir before it will flow.[2]

The need for often complex and expensive production techniques means that while bitumen and heavy oil reserves are vast, only a tiny fraction of either resource will ever see the inside of a refinery.

BITUMEN: THE EARLY RECORD

Awareness of the oil sands deposits in Canada predates recorded North American history, and almost certainly occurred several thousand years ago. Natives knew about bitumen from the time they began to navigate the Athabasca and Clearwater Rivers, where the oil sands lie exposed for long distances.

The first recorded mention of Canada's bitumen deposits goes back to June 12, 1719. According to an entry in the York Factory journal, on that day Cree Indian Wa-Pa-Sun brought a sample of oil sand to Henry Kelsey of the Hudson's Bay Company.[3] When fur trader Peter Pond travelled down the Clearwater River to Athabasca in 1778, he saw the deposits and wrote of "springs of bitumen that flow along the ground." A decade later, Alexander Mackenzie saw Chipewyan Indians using oil from the oil sands to caulk their canoes.

The first commercial effort to exploit the oil sands came as a result of John Macoun's 1875 reconnaissance survey for the Geological Survey of Canada. His report (and later reports by Robert Bell and R.G. McConnell) led to test holes in the oil sands. The Survey probably hoped to find free oil at the base of the sands, as drillers had in the gum beds of southern Ontario a few decades earlier. Although the Survey's three wells failed to find oil, the second was noteworthy.

Drilled at a site called Pelican Portage, the well blew out at 235 metres after encountering a high-pressure gas zone. According to drilling contractor A.W. Fraser:

> *The roar of the gas could be heard for three miles or more. Soon it had completely dried the hole, and was blowing a cloud of dust fifty feet into the air. Small nodules of iron pyrites, about the size of a walnut, were blown out of the hole with incredible velocity. We could not see them going, but could hear them crack against the top of the derrick.... There was danger that the men would be killed if struck by these missiles.*[4]

Fraser's crew unsuccessfully tried to kill the well by casing it, then abandoned the well for that year. They returned in 1898 to finish the job, but again they failed. In the end, they simply left the well blowing wild. Natural gas flowed from the well at a rate of some 8 500 000 cubic feet per day until 1918. In that year a crew led by geologist S.E. Slipper and oilman C.W. Dingman finally shut in the well.[5]

While the Pelican well was blowing wild, others drilled on the Athabasca deposit in hopes of finding free oil but met with no success. One of the more colorful early promoters was Alfred von Hamerstein, who claimed to be an immigrant German count and supposedly first saw the oil sands deposit on his way to the Yukon.

In 1907 Von Hamerstein made a celebrated presentation to a Senate committee investigating the potential of the oil sands. "I have all my money put into it [the Athabasca oil sands]," he said, "and there is other peoples' money in it, and I have to be loyal. As to whether you can get petroleum in merchantable quantities . . . I have been taking in machinery for about three years. Last year I placed about $50 000 worth of machinery in there. I have not brought it in for ornamental purposes, although it does look nice and home-like."[6] Recent scholarship, however, suggests that wild speculation and fraud marked his ventures.[7]

In 1913, the federal Department of Mines assigned Dr. S.C. Ells, an engineer, to investigate the economic potential of the oil sands around Fort McMurray. Ells' studies convinced him that the oil existed only in combination with sand. He came up with the idea of using it for road-paving, and used it to lay several sections of sidewalk and road surface in Edmonton. In 1925-26, three kilometres of highway in Jasper Park also received a pavement of McMurray asphalt. Road paving with Athabasca oil sand became a cottage industry in Alberta during the early part of the century, serving as pavement as far east as Ottawa and Petrolia, Ontario. However, this use was basically uneconomical.

The Alberta Research Council, which also had a keen interest in the oil sands, assigned Dr. Karl Clark to research them. He and his associate, Sid Blair, concluded that oil sands had more potential as a source of oil than as paving material. They studied various means of separating the oil from the sand, as Ells had. In 1923, the two built the first bench model of a hot-water separation plant at the University of Alberta. They constructed a larger plant in railway yards near Edmonton in 1924, and the Research Council later constructed a pilot plant on the Clearwater River near Waterways. Both ventures employed Clark's hot-water separation process.

These efforts led to an enterprise that is unique to the Canadian petroleum industry: Extracting oil from open pit mines. Ever since Clark and Blair developed their first experimental plant, the industry's large-scale efforts in the Athabasca deposit have focused on mining the oil sands. This activity is possible near Fort McMurray because the deposits of oil sand ore are rich, thick and close to the surface.

The oil sands plants perform three processes. First, they mine the oil sand ore. Second, they separate bitumen from the sand which encases it. Third, they upgrade the bitumen, turning the tar-like bitumen into a light oil suitable for refining.[8] Nowhere but around Fort McMurray do oil sands plants like these exist.

Today's oil sands behemoths are successors to entrepreneurial efforts to retrieve oil from the oil sands in the 1920s and 1930s. Notable among these were Abasand Oils Limited and the International Bitumen Company.

Two American promoters, Max Ball and B.O. Jones of Denver, entered the oil sands scene in 1930 when they organized Abasand. They bought the Alberta Research Council's Fort McMurray plant and began building their own plant west of Waterways in 1935. Although scheduled to go into operation by September 1936, forest fires and equipment supply delays held up plant construction. Mining at the Abasand plant finally began in 1941. In its first four months of operation, the plant processed 18 475 tonnes of oil sand to produce 17 000 barrels of oil. But in November 1941, fire destroyed the plant.

The company rebuilt the plant and in 1943 the federal government took it over as part of the war effort. Federal researchers felt the hot water process was too costly because of heat loss and started to experiment with a cold water process. Work at the plant ended in 1945 when another fire destroyed the entire operation.

The International Bitumen plant actually preceded Abasand, but its active life was somewhat longer. Founded by R.C. Fitzsimmons in 1925, International Bitumen used a combination hot water and solvent method to produce pure bitumen at a location called Bitumount on the Athabasca River. The plant operated for several years but eventually faltered as a commercial enterprise. However, it came back to life during World War II.

Frustrated with the problems and delays at the Abasand plant, in 1943 the Alberta government made plans to build its own oil sands plant at the Bitumount site. Although construction dragged on until well after the war, this plant which cost $725 000 (one third the cost of the Abasand plant) went on production in 1948. Tests during its first year and during 1949 were successful, but the plant then closed. Then, with the oil finds at Leduc in 1947, the flurry of activity in the oil sands which climaxed during World War II dragged to a halt after the Leduc oil finds of 1947.

THE OIL SANDS AFTER LEDUC

The Leduc oil strike and the many other petroleum finds that quickly followed naturally drew industry attention away from the oil sands. Bitumen from the Athabasca oil sands could not compete with inexpensively-produced conventional light oil. Though interest wavered in those years, it did not die.

Oil sands pioneer S.M. Blair (who had worked with Karl Clark to develop the hot water separation process) suggested that the Alberta government assess the potential for development in the oil sands and was hired for the task. Published in 1950, the Blair report stated that oil sands development could be economic for projects producing 20 000 barrels or more of oil per day. He envisioned such a plant costing $43 million and generating a five to six percent annual return on investment. He believed that such an operation could profit in a market where conventional oil is fetching only $2.70 per barrel because synthetic oil was such an attractive feedstock.[9] Far more valuable refined products can come from a barrel of synthetic than from a barrel of conventional oil.

In 1951, Alberta sponsored an oil sands conference to discuss oil sands geology, mining, recovery, transportation and refining. Nathan Tanner, Alberta's Minister of Mines and Minerals, outlined provincial policy on oil sands leasing and royalties. Many oil companies participated in the conference. And shortly after its conclusion, a dozen took out exploration permits covering 20 000 hectares apiece.

The provincial policy allowed companies to take out "prospecting permits" for up to 20 000 hectares for a one-year period, renewable for up to two additional years. Companies could then lease enough of the permit area to support an oil sands plant. The remainder of the exploration area reverted to the province.

None of the companies with permits proceeded immediately to the lease stage. The main reason was a government stipulation that the lessee had to begin construction of a commercial plant within two years of receiving the lease, and begin operating the plant within five.[10] The government later changed the rule so the lessee's obligation was to build within a year of receiving a government order to do so. The government pledged to be reasonable in making such requests, and several companies took out leases.[11]

During the late 1940s and the 1950s the oil industry learned much about the oil sands. Drilling and other geological work defined three major deposits. Industrial engineers and chemists improved processes for extracting and upgrading the bitumen. And the petroleum community in general came to a better understanding of the business potential of the oil sands as an energy resource.

With knowledge came action. In 1959, Cities Service Athabasca Inc. constructed a 3 000 barrel per day plant at Mildred Lake, north of Fort McMurray. This plant extracted bitumen at a field facility, then shipped the raw bitumen to a pilot refinery.

The forerunner of Syncrude, this was the first oil sands project to use a bucketwheel mining machine. Fitted with replaceable teeth, this rotating wheel of buckets constantly scooped away chunks of the oil sand ore. Some 4.3 metres wide, 20 metres long and 60 tonnes in weight, this bucketwheel used closed-circuit television to allow the operator to watch the discharge conveyor. The electrically powered machine could mine 200 tonnes of oil sand per hour.[12]

GREAT CANADIAN OIL SANDS LTD.

In 1962, Great Canadian Oil Sands (GCOS) Limited received approval from the Alberta government to build and operate a 30 000 barrel per day plant near Fort McMurray. The plant was to produce 240 tonnes of sulphur and 900 tonnes of coke per day as by-products.[13] Because at that time the industry was having difficulties marketing its oil, the provincial government established a policy to limit oil sands production. According to this policy, synthetic oil from the oil sands could supplement conventional oil sales, but could not displace it. Oil from the GCOS plant could not exceed five per cent of total volumes in markets already supplied by conventional Alberta oil.

Financial difficulties delayed construction of the GCOS plant until a new investor, Sun Oil Company's Canadian subsidiary, (today known as Suncor) was found. The capacity of the proposed plant increased to 45 000 barrels per day and the cost escalated from $122 to $190 million. The larger plant received approval in 1964 and went into commercial production in September 1967.[14] The final cost: $250 million.

During the opening ceremonies for the plant, Sun Oil Company chairman John Howard Pew (then 85 years old) made remarks which still ring true:

> *No nation can long be secure in this atomic age unless it be amply supplied with petroleum.... It is the considered opinion of our group that if the North American continent is to produce the oil to meet its requirements in the years ahead, oil from the Athabasca area must of necessity play an important role.*[15]

The Suncor plant was a landmark in oil sands development. It pioneered technology for bitumen extraction and upgrading, and it was the world's first large-scale commercial plant. In the early years it was not particularly profitable, but the plant was nonetheless able to cover operating expenses from the sale of its own production. And in 1979, when federal policy permitted the company to charge world price for its oil, the plant finally became a money-making asset for Suncor.

Since construction, Suncor has progressively increased the efficiency, reliability and capacity of its oil sands plant. As a result, after its first quarter-century of operation this oil sands pioneer had increased capacity by more than one third. During 1991 its average daily production was 60 600 barrels and its cost of producing synthetic oil was $15.75 per barrel. Compared to the cost of producing conventional oil, that cost is high. But since oil sands development does not involve risky and expensive exploration, synthetic oil can be competitive with other forms of oil production. The Suncor plant offers bottom line proof.

SYNCRUDE

In 1962 (the same year the GCOS proposal went up for approval) Cities Service Athabasca Inc. proposed a 100 000 barrel per day plant at the site of its Mildred Lake pilot project. Including a pipeline to Edmonton, the plant was to cost $56 million, with construction beginning in 1965 and completing in 1968.

However, the Oil and Gas Conservation Board had concerns about competition between synthetic oil and conventional oil for limited markets. It therefore decided not to bring too many oil sands plants on stream at once, and rejected the Cities Service proposal in favor of the GCOS project.

Cities Service later reapplied for a much larger plant, and the proposal received approval in late 1969. The Syncrude plant which resulted went into production exactly two centuries after Peter Pond's first siting of the oil sands in 1778. But before the plant shipped its first barrel of oil, the project went through many trials.

The reason for the long gap between approval and completion was an alarming escalation of costs that beset all major North American projects in the 1970s. High inflation multiplied budgets for practically every aspect of the Syncrude project.

Reviewing project costs in late 1973, the Syncrude consortium found that costs had more than doubled, from $1 billion to $2.3 billion. In December 1974, Atlantic Richfield (whose American parent needed cash to develop its Prudhoe Bay interests) withdrew its 30 percent participation in the project. A few days later, the three remaining partners informed the Alberta government that the maximum risk they were willing to take on the project was $1 billion. They would need to find another $1 billion of risk capital if the project were to go on.

By this time the world was in the thralls of an energy crisis. Beginning in 1973, the members of the Organization of Petroleum Exporting Countries (OPEC) had taken advantage of tight world oil supplies to rapidly and regularly increase prices. Policy makers in the oil consuming countries therefore considered it a matter of national urgency to develop stable, secure energy supplies. Because the resource was so large and development was clearly possible, the oil sands looked like Canada's best bet. As a result, the prospect that the Syncrude project would collapse was a matter of both political and economic concern.

Alberta invited the other governments of Canada to participate as commercial partners in the project. The province also reviewed the cost estimate given by the oil companies. When it found that the consortium's cost estimates were not out of line, Alberta helped convene a historic meeting in Winnipeg in February, 1975. That meeting salvaged the project.

The federal government took a 15 percent interest, Alberta 10 percent and Ontario five percent. The private partners agreed to take a $1.4 billion interest in the project, but gave Alberta the option to convert a $200 million loan to Gulf and Cities Service into ownership interests. Alberta also took full ownership in the no-risk pipeline and electrical utility which the plant needed.[16]

The plant went into operation in the summer of 1978 and produced 5 million barrels of oil within a year. World oil prices leaped skyward in 1979-80 and remained high for the first half of the 1980s. This helped Syncrude become successful financially as well as technically.

OTHER MINING PROJECTS

Although two additional mining projects received serious consideration, both fell victim to the declining fortunes of the petroleum industry after 1980.

In 1962 – the year of the original GCOS and Syncrude proposals – Shell Oil of Canada (today Shell Canada) also proposed a plant near Fort McMurray. The most ambitious of the original proposals, Shell called for a 130 000 barrel per day, $50 million plant. Production would begin in 1969. However, escalating costs and Alberta's policy of restricting oil sands development led Shell to withdraw the application.

With the discovery of Prudhoe Bay oil in 1968, the decision not to invest heavily in the oil sands seemed prudent. The Alaskan discovery was immense, and there was a chance of similar finds in the Canadian Arctic. Like most other companies, Shell came to question the wisdom of investing in high-cost synthetic oil.

World oil supplies and the pricing outlook changed drastically during the 1970s, however, and in 1979 a consortium led by Shell revised and resubmitted the Alsands proposal to the Energy Resources Conservation Board. By this time projected to cost $6 to 7 billion, the new project would have been larger than Syncrude. But by 1982, political wrangling between Alberta and the federal government had delayed construction of the Alsands plant. Costs had doubled, and signs of an impending oil surplus had begun to appear. Partners began to pull out of the consortium. Despite a last-ditch effort by the Alberta and federal governments to save Alsands, the economic window had closed. The project went on the shelf.

Yet another proposal to develop the oil sands came in 1988, when another consortium proposed construction of a mining project. Named the OSLO project (for Other Six Lease Operations), the $4.1 billion project would have included a mine and extraction plant 80 kilometres north of Fort McMurray. However, its upgrader would have been at Redwater, 60 kilometres northeast of Edmonton. Production would have totalled 77 000 barrels per day. Construction was to begin in 1993, with the project going on stream in 1996. However, in 1991 the project succumbed to a growing financial crisis within the industry.

Although the Alsands and OSLO projects failed, Suncor and Syncrude have made Canada the world's largest producer of synthetic oil. Those two plants produce about 16 percent of Canada's light and medium oil. By closing the gap between conventional and synthetic oil production costs, they have boosted the economics of Alberta's immense deposits of oil-bearing sands. And they have contributed in countless ways to Canada's economy. Perhaps nothing sums up Canada's oil sands achievements and potential better than the Syncrude operation.

Syncrude is one of the most complex industrial operations in the world, processing 330 000 tonnes of oil sand ore per day – more than enough to fill four football stadiums. And it synthesizes nearly 65 million barrels of light oil per year, an amount equal to almost 12 percent of Canada's oil requirements.

Syncrude operates the largest mine in the world, with 4 500 permanent employees and usually an additional 1 000 to 1 500 contract personnel on site. In Edmonton, 130 employees spearhead the company's research and development activities, which cost about $20 million per year. So successful has been Syncrude's development and application of new systems and technology that its production costs per unit of oil declined by more than half in the decade ending in 1991.

Early in that year the plant delivered its 500 millionth barrel of oil. Although a few Canadian oil fields have produced that much oil, none have done so in such a short time, and none have done so without depletion of the field. Despite its enormous achievement, Syncrude still has vast resources for production.[17]

UNDERGROUND RECOVERY

The oil sands projects mentioned above were all mine-based, exploiting bitumen in the Athabasca oil sands near Fort McMurray from open-pit mines. After years of experimenting with ways to recover bitumen from deeper deposits, the industry has discovered that the only way to develop these resources successfully is through in situ production.

"In situ" means "in place" and refers to recovery techniques which apply heat or solvents to oil reservoirs beneath the earth. The first in situ experiment in Alberta took place in 1910, when a Pittsburg-based outfit, the Barber Asphalt and Paving Company, drilled a bore hole into bitumen and pumped in steam to liquefy the oil. The experiment failed. In the early 1920s other in situ experiments also took place, but none were commercially successful.

In the mid-1920s, a remarkable and persistent experimenter named Jacob Owen Absher incorporated the Bituminous Sand Extraction Company. Absher began his in situ experiments in 1926, and carried on numerous experiments over the following five years – efforts that drew the interest of oil sands pioneers Sidney Ells and Karl Clark. Absher not only used steam to melt the bitumen, but also tried igniting fires within his wells. In the end, however, Absher was unable to produce oil from the sands.

Today, there are several varieties of in situ technique, but the ones which work best in oil sands use heat. Commercial projects in Alberta pipe high-pressure steam into the oil sands reservoir. Other experimental projects actually ignite the oil underground, then pump air below the surface to keep combustion going. These techniques effectively melt the oil, which is pumped to the surface.

The most dramatic proposal for an in situ project came from Richfield Oil Company. In 1959 Richfield suggested an experimental plan to release liquid hydrocarbons from the sand with an underground nuclear explosion. The company proposed detonating a nine-kiloton explosive device below the oil sands at a site 100 kilometres south of Fort McMurray.[18] Thermonuclear heat would liquefy the oil, enabling the company to produce it.

This proposal came remarkably close to happening. The project received Canadian provincial and federal approval,[19] and the United States Atomic Energy Commission agreed to provide the device. But before the experiment could take place, public pressure for an international ban on nuclear testing had mounted. This created a climate which killed the plan. Eventually, other in situ plans went ahead.

ALBERTA'S OIL SANDS DEPOSITS

As companies drill, they keep records of bitumen and other oil and gas shows. Geologists use these findings to map underground deposits. A geological report first defined the important oil sands deposit surrounding Cold Lake in eastern central Alberta in 1961. This discovery was largely a result of exploration by Imperial Oil. The drilling process has also helped define the Athabasca, Wabasca, Peace River and other less significant oil sands deposits in Alberta.

The first companies to experiment in Alberta's deeper oil sands deposits included Shell Canada, Imperial Oil and BP Canada, each of which began in situ work in the 1960s. During the 1970s other companies also set up oil sands projects, and in situ facilities dotted the maps of Alberta by the end of the decade.

In the mid-1980s, in situ pilot plants gave way to commercial projects. Development went ahead in those years mainly for three reasons:
1. The industry had enjoyed technical successes in its pilot projects, and relatively small-scale plants were less risky and quicker to bring on stream than megaprojects.
2. Alberta and the federal government had devised a fiscal regime to encourage in situ development.[20]
3. Oil prices were still high.

Together, these factors led to a brief but dramatic flurry of oil sands activity. There have been pilot projects in all of the oil sands deposits, but the larger projects – for example, at Wolf Lake, Elk Point and Lindbergh – exploit the Cold Lake deposit.

The Peace River Deposit

Shell began drilling in the Peace River area in 1941, and continued drilling there until 1947. Its efforts confirmed that bitumen was present in widespread deposits. The company began in situ experiments there in the early 1960s, ended them in 1966 and reactivated them in 1972. The company established a pilot project at its Peace River site in 1979. In 1984, Shell announced that it would expand this pilot into a small-scale commercial plant at a cost of $200 million. The Peace River complex, which included a steam-driven production process and plans for 212 wells, came on stream in October 1986, with production at 10 000 barrels per day.

The Cold Lake Deposit

After the key Athabasca oil sands deposit, the most important of these vast resources is the deposit around Cold Lake. Imperial Oil acquired leases for this area in the 1950s, and eventually estimated bitumen in place on those leases at 44 billion barrels. As thick as a 20-storey building is tall, the oil sands lie about 500 metres below ground. In 1964, Imperial set up a four-well pilot plant at Ethel Lake. Eight years later, the company developed a 23-well pilot at May Lake. And in 1975 the company began a 56-well project at Lemming. That plan was so successful the company later increased the number of operating wells to 300.[21]

Encouraged by these successes, in 1979 Imperial applied to build a $7 billion, 135 000 barrel per day in situ plant and upgrader at Cold Lake.

By 1982, that proposal had gone the way of Alsands. The National Energy Program, with its array of taxes and incentives, had led to political conflicts between Edmonton and Ottawa, and Imperial was unable to arrive at a satisfactory agreement on fiscal terms with the governments. In addition, project costs had soared, and world oil prices were beginning to drop. These factors made the project extremely risky, especially for a company operating alone. Under this collective burden of unfavorable conditions, Imperial shelved the project.

In the mid-1980s, the company revived the Cold Lake project in much less ambitious form. Constructed in stages, the project will eventually include 16 separate phases of development, each of which may require as many as 800 wells over 25 years.

The first six phases went on stream in 1986, and the company let economic conditions dictate the construction schedule for the project's additional phases. Because of poor economic conditions within the oil industry generally, the company did not develop additional phases as soon as originally planned. By 1991, the project's average daily bitumen production from the first six phases was 90 000 barrels.

Imperial's modular approach to this project enabled a version of its proposed megaproject to go ahead with less risk to the company. The operation's focus was production: to make the project affordable, Imperial eliminated the upgrader from its plans. Rather than upgrade production into synthetic oil, the plant mixes bitumen with lighter hydrocarbons, known as diluents. Pipelines then carry this thinned-out oil to refineries capable of using it.

Although the Imperial plant is the largest, by 1992 there were perhaps 30 in situ operations in Alberta. Together, they produced nearly 125 000 barrels of bitumen per day, or about 7.5 per cent of Canadian oil production. The value of that oil – marketed mostly in the United States, but also in European and Pacific Rim countries – was some $400 million per year.

Underground Mines

A twist on the in situ approach uses an underground mine to produce bitumen. The Alberta Oil Sands Technology and Research Authority, a provincial agency set up in 1974, began investigating the idea of using a system of shafts and tunnels in the late 1970s.

The result of its investigation is the Underground Test Facility, a bitumen mine 60 kilometres northwest of Fort McMurray. Several private sector partners have interests in the project, as does the federal Department of Energy Mines and Resources.

The mine is in a limestone formation underneath an oil sand reservoir. Constructed in 1984, this underground venue provides the drill site for horizontal wells which reach upward into the sands. Steam piped through one set of wells heats the bitumen, which drains down another set of wells into production facilities in the mine. Although experimental, project results have exceeded the results predicted by computer modelling. And the project has greatly extended the technology available to develop the oil sands.[22] It is not yet ready to become a commercial technology, however.

HEAVY OIL

Heavy oil is a sister resource to bitumen, but lighter. Reservoirs are much smaller than the oil sands deposits but similar in the sense that only a small percentage of the crude oil is recoverable.

Often called "conventional heavy oil," this low-density oil can be recovered by conventional drilling techniques or by waterflood, a technique of injecting water into the reservoir to increase pressure, thus forcing the oil toward the well bore. But as heavy oil can be quite viscous, it sometimes requires some form of heat or solvent and pressure before it will flow into a well bore. When heavy oil requires these techniques to go into production, it is known as "non-conventional heavy oil."

The first heavy oil discoveries came with the pursuit of conventional light and medium crude. Because much of western Canada's heavy oil lies close to the surface, early explorers using older rigs discovered many of those pools before they came upon the deeper light oil reservoirs.

One of the first finds was in the Ribstone area near Wainwright Alberta in 1914. The province's first significant production of heavy oil, almost 6 000 barrels in 1926, came from the Wainwright field. A small-scale local refinery distilled the heavy goo into usable products.

Elsewhere in Alberta, petroleum explorers made other heavy oil finds as they pursued the elusive successor to the Turner Valley oil field. Many of these fields produced only small volumes. The recovery techniques of the day, combined with the low price of oil and the nature and size of the finds, meant that most of the oil remained undeveloped.

The most important exception was at Lloydminster. While the first discovery occurred in 1938, serious development did not begin until Husky Oil moved into the area after World War II.

Husky Oil was born during the Depression through the efforts of Glenn Nielson, an Alberta farmer driven to bankruptcy. Nielson had moved to Cody, Wyoming, by the time he founded Husky as a refining operation. He turned his attention back to Canada after the Second World War, deciding to set up a refinery at Lloydminster. Steel was scarce, so Husky dismantled a small Wyoming refinery which provided bunker fuel to the American Navy during the war, loaded the pieces onto 40 gondola cars and shipped them north by railway.

The company reassembled the 2 500 barrel-per-day facility in 1946, and the refinery went into production the following year. Strategically located between the Canadian Pacific and Canadian National railroad tracks in Lloydminster, the refinery soon began to get contracts for locomotive bunker fuel. Husky also found a strong market in asphalt for road building.

Husky's move into the area spurred drilling and production. Within two years of Husky's arrival, there were storage space shortages. Producers solved the problem by storing the oil in earthen pits holding up to 100 000 barrels each. For a while Husky bought the oil by weight rather than volume since it was clogged with earth, tumbleweed and jackrabbits. The company had to strain and remeasure the crude oil before it could begin refining.

Husky began producing heavy oil from local fields in 1946 and by the 1960s was easily the biggest regional producer. In 1963 the company undertook another in a series of refinery expansions to 12 000 barrels per day. To take advantage of the growing market for Canadian oil, Husky developed a project to deliver heavy oil to national and export markets.

The key to the $35 million project was the construction of a reversible pipeline which could move the viscous heavy oil into the marketplace. The completed 116-kilometre "yo-yo" pipeline – the first in the world – brought condensate from the Interprovincial Pipe Line station at Hardisty, Alberta. Husky mixed this very light hydrocarbon with heavy oil, enabling it to flow more easily. The company then pumped the blend back through the pipeline (hence the nickname "yo-yo") to Hardisty. From there Interprovincial Pipe Line took the heavy oil eastward to market.[23]

These developments made heavy oil more than a marginal resource for the first time. Within five years, area production had increased five-fold to 11 000 barrels per day. By the early 1990s, production from the heavy oil belt from all players there was some 250 000 barrels per day. Husky is still one of Canada's biggest heavy oil producers.

UPGRADERS

As Canada's conventional oil reserves decline, the industry's attention is turning increasingly to heavy oil and bitumen. Enhanced recovery techniques now bring a higher percentage of reservoir oil to the surface. Research and development technologies have increased the amount producers can extract. Even small improvements in technology applied to such huge resources could mean enormous additions to Canada's recoverable crude oil reserves.

Few Canadian refineries can process more than small amounts of heavy oil, so heavy oil has traditionally gone to United States asphalt plants for upgrading. In the late 1970s, a group of heavy oil producers (Gulf, Husky, Shell, Petro-Canada and SaskOil) proposed the Plains Upgrader. This facility would have cost $1.2 billion and upgraded 50 000 barrels of heavy oil per day. Gradually, however, consortium members pulled out of the project as they concluded that the high cost of upgrading would make the project uneconomical. In the end, only the crown corporations, Petro-Canada and Saskoil, remained.

While the Plains partnership collapsed, the idea survived.

Their partners gone, the Saskatchewan Crown corporation (SaskOil) suggested reducing upgrader costs by integrating with the Consumers' Co-operative Refinery in Regina. This would eliminate duplication in facilities and infrastructure by taking advantage of existing land, processing units, storage and pipeline facilities, technical and operating staff and management.

The Co-op refinery is a product of the Co-operative movement, which began in Britain in the mid-1800s. This movement consists of co-operative associations which buy in bulk at discount rates, refunding the money they save to their members, who are also owners. The co-op movement has been particularly successful in western Canada, especially in Saskatchewan where the agricultural based retailer dominates.

During the first years of the Depression, farmers formed co-operatives to buy petroleum products at wholesale prices. Because the spread between the wholesale and retail price of farm fuels was quite wide, these organizations were initially successful. But between 1931 and 1934, British American (later Gulf Canada) and Imperial Oil bought three independent regional oil refineries which served southern Saskatchewan, reducing the number of local suppliers from four to two,[24] and in 1933 the federal government imposed a protective duty of 3.7 cents per gallon on imported gasoline.[25]

Without competition among refiners, co-ops were not able to purchase bulk gasoline and lubricating oils at greatly reduced wholesale prices. So a feisty group of Regina-area co-operatives decided to create a refinery co-op, and raised $32 000 to construct a 500 barrel-per-day skimming plant.

This tiny refinery could do no more than distill light oil shipped in by rail from the United States, but it was the world's first co-operative refinery. And it was phenomenally successful: from start-up in mid-1935, co-op members saved more than $28 000 in the first six months.[26]

The Co-op plant was a small but modern refinery when talk about a refinery/upgrader complex began in the early 1980s. Both the federal and Saskatchewan governments had forbidden their Crown corporations to participate in the project, yet both governments took part themselves. The province had a particular interest, since an upgrader would increase the market for heavy oil from Saskatchewan's fields. This would give the provincial oil industry an important boost. The federal government saw the project as an opportunity to move the nation one small step towards the stated goal of crude oil self-sufficiency. For its part, the Co-op wanted an assured supply of readily refinable crude oil for its refinery.

Accordingly, Saskatchewan took a 20 percent equity position and guaranteed loans equal to 45 percent of the project. In exchange, it became a 50 percent partner in the combined operation with Consumers' Co-op, which committed its existing refinery (valued at $500 million) to the project. The federal government guaranteed loans equal to 35 percent of the project. Repayment on the principal of the loans would not begin until late 1992.

On stream in 1988, the Consumers' Co-op refinery/upgrader complex is a 50 000 barrel per day facility. The $700 million upgrader provides upgraded oil as refinery feedstock. Since refinery demand fluctuates, the upgrader usually upgrades more oil than the co-op can use. The refinery complex delivers the surplus, on average 15 000 barrels per day, into the IPL system for transport to markets in central Canada and the United States.[27]

Appropriately, the company with the most extensive experience in the heavy oil belt was the one to propose and eventually develop Canada's other heavy oil upgrader. Husky began to prepare for the upgrader by building a new 25 000 barrel-per-day refinery next to the old plant. This refinery, which processes heavy oil into asphalt and provides light oils for use in the upgrader, replaced the old one in 1983.

After a series of false starts, in 1988 Husky and its three partners announced a firm agreement to construct the Bi-Provincial Upgrader. Located just east of Lloydminster, this $1.6 billion upgrader received most of its funding from government. The federal, Alberta and Saskatchewan governments own 31.67 percent, 24.16 percent and 17.5 percent respectively. The balance belongs to Husky.

Under the terms of the original agreement, Husky would receive 50 per cent of the plant's net revenue plus a 10 per cent return on investment until the investment was recovered. The balance of plant profit would go proportionally to Husky's partners. A wrinkle in this arrangement occurred as the project neared completion, however, when Saskatchewan's newly-installed NDP government refused to pay its share of $190 million in cost overruns. The other players eventually agreed to pay Saskatchewan's share, but would withhold returns to that province until they had recovered Saskatchewan's $33 million in arrears.

The upgrader went on stream in mid-1992, but would probably require a year or more of unplugging the bottleneck before it could reach productive capacity of 46 000 barrels per day. The plant upgrades Lloydminster area heavy oil and Cold Lake bitumen, making still more of those resources available for central Canadian and American markets.

Large cost discrepancies exist between the Bi-Provincial upgrader ($1.6 billion for 46 000 barrels per day capacity) and the Co-op upgrader ($600 million for 50 000 barrels per day.) This is due to the heavier grades of oil the Bi-Provincial facility processes. However, its output is far more desirable than the Co-op's. This critical difference means the Bi-provincial pays less for its feedstock and receives more for its output than does the Co-op plant. From the beginning, forecasts about these differentials were vital factors in economic calculations for the two projects.

THE ECONOMICS OF DEVELOPMENT

Three quarters of a century of experience in Canada's oil sands have solved many of the associated technological challenges. But economics and policy will largely determine future development. Experience in recent history illustrates how sensitive these unconventional resources are to political and economic factors.

The oil sands in particular have been politically sensitive since the 1920s, when Alberta saw its future tied to the potential of the resource. In the 1970s, the oil price spiral focused national attention on these enormous deposits. Suddenly, the oil sands' time had apparently arrived.

Syncrude went on stream and began to prosper, and dots designating experimental oil sands projects soon peppered the maps of Alberta deposits. Heavy oil production began to surge. But as project proposals multiplied, delays in government approvals (coupled with rapid escalations in construction costs) diminished their allure. Then worldwide oil shortages transmuted into a global glut of the stuff. Oil prices began to fall. And the oil sands almost completely lost their lustre. The industry began reducing its investment in non-conventional resources.

Governments took on the role of heavy oil and oil sands advocate as project economics deteriorated. There were two reasons for this. First, heavy oil and oil sands projects are capital intensive and create economic spin-offs, which generate economic growth and tax revenue. Second, governments took seriously the importance of oil as a strategically important commodity.

Reflecting these realities, new commercial oil sands developments flourished briefly in the mid-1980s, but relied on low royalties and government incentives to proceed. Both heavy oil upgraders also required government participation. Both of these risky projects are industry/government partnerships in which taxpayers put up most of the cash.

The prospects for further development diminished in early 1986 when a precipitous collapse in oil prices once again threatened oil sands development and other high-cost energy projects. Characterized by lower oil prices, reduced industry income, low returns on petroleum investment and market-oriented energy policies, the economic environment of the 1990s is tough.

Until circumstances change, it will be difficult for the petroleum industry to develop western Canada's vast bitumen and heavy oil resources.

Chapter 9

Frontiers of Muskeg, Ice and Water

Geophysical observer using a gravity meter in the Mackenzie Delta on a survey for British American Oil Company, called the Reindeer Project, 1965. (Glenbow Archives/NA-2864-41976)

Canada's early petroleum discoveries happened near population centres or along lines of penetration into the northern frontier. The first oil play was in southern Ontario; the first western natural gas discovery occurred on a Canadian Pacific Railways right-of-way; the site of the first discovery in the far north was along the Mackenzie River, the great transportation corridor into Canada's Arctic. From those haphazard beginnings the search for petroleum eventually spread to the fringes of Canada and beyond, to the ocean's continental shelves.

THE NEAR NORTH

In oil industry terms, "frontier" usually refers to Canada's offshore or Arctic regions. However, the early frontiers were comprised of Canada's muskeg belt in the Northwest Territories and the Yukon.

In those areas, the climatic rigors of winter give way to fearsome logistical problems when summer temperatures melt the top frost. Unfrozen muskeg could swallow a caterpillar tractor whole. Roads constructed during the winter disappeared into deep oblivion.This landscape dictated two approaches. Earlier, sites for wildcat wells were only accessible by river; later, as drilling intensified and technology developed, winter was chosen as the drilling season, despite the cold. Summer exploration on muskeg is a problem that challenges the petroleum industry to this day and that has led to innovative solutions. An expensive but effective way to drill in muskeg year-round is to drill wet sites in winter when they are frozen, saving the drier sites for summer. During the late 1970s and early 1980s, helicopters sometimes flew dismantled rigs from site to site.

Before the turn of the century, employees of the Geological Survey of Canada canoed and tramped the north, mapping the northern extension of the Western Canada Basin. These surveys inspired many early wildcat wells.

Speaking before an 1888 Senate committee, former Hudson's Bay Company Chief Factor William Christie saw little difficulty in drilling for oil in the north. By the standards of his company, the work was hardly daunting.

In 1893-4, the Dominion government entered the fur trade region by drilling a well on the Athabasca River. Although the well was unsuccessful, it was clear that conveying drilling equipment into the north by boat and drilling along the rivers was practical.

In the exploration boom of the 1950s, the industry encountered muskeg, cold and the absence of roads.

But gradually exploration pushed the frontier back into northern Alberta, northeastern B.C. and across the 60th parallel into the Yukon and Northwest Territories. Those efforts rewarded corporate frontiersmen with such major oil finds as Wizard Lake, Bonnie Glen and Swan Hills.

ABOVE THE ARCTIC CIRCLE

Beyond the Arctic Circle lie the lands and waters of the midnight sun. In those barren areas, the sun stays up 24 hours per day in summer but does not rise in winter.

The definitive push north of the Arctic Circle took place in 1957 when Western Minerals and a small exploration company called Peel Plateau Exploration drilled the first well in the Yukon. In order to equip and supply the well, some 800 kilometres from Whitehorse at Eagle Plains, Peel Plateau hauled 2 600 tons of equipment and supplies by tractor train. This achievement involved eight tractors and 40 sleighs per train, for a total of seven round trips.[1] Drilling continued in 1958, but the company eventually declared the Peel Plateau well dry and abandoned.

Stirrings of interest in the Arctic Islands as a possible site of petroleum reserves came as a result of "Operation Franklin," a study of Arctic geology directed by Yves Fortier under the auspices of the Geological Survey of Canada. This and other surveys confirmed the presence of thick layers of sediment containing a variety of possible hydrocarbon traps.

The petroleum industry applied to the federal government for permission to explore these remote federal lands in 1959, before the government had begun regulating such exploration. The immediate result was delay. But in 1960, the Diefenbaker government passed regulations, then granted exploration permits for 16 million hectares of northern land. These permits granted mineral rights to companies, on condition they spend a certain amount of money or carry out a certain amount of exploration within a specified time.

The first well in the Arctic Islands was the Winter Harbor #1 well on Melville Island, drilled in the winter of 1961-62 by Dome Petroleum. Equipment and supplies for drilling and for the 35-man camp came in by ship from Montreal. Although this well was dry, as were two others drilled over the next two years on Cornwallis and Bathurst islands, all three were technical successes. There was no doubt that high Arctic drilling was possible.

Although these pioneering wells were milestones in frontier development, between 1964 and 1967 exploration in the far north all but stopped. Arctic exploration was proving to be extremely expensive and risky, and exploration prospects in less remote areas of the world were more attractive than in the Canadian north. For example, the Mitsue, Nipisi and Rainbow plays kept many investors in Alberta. Outside the country, offshore California and the North Sea were strong magnets for petroleum investment.

The federal government's eagerness to encourage Arctic Islands exploration, partly to assert Canadian sovereignty, led to the formation of Panarctic Oils in 1968. Panarctic consolidated the interests of 75 companies and individuals with Arctic Islands land holdings plus the federal government as the major shareholder.

The company had a long and complicated birth. When the deal was complete in 1968, the federal government held 45 per cent of the new company's equity. Other than a brief period of involvement during World War II, Panarctic marked the federal government's first direct and permanent entry into the oil and gas business. Since its formation, the company has been the principal oil and gas operator in the Arctic Islands.

In that role it has spent some $900 million and has been the operator for roughly three-fourths of 175 wells or more drilled in the high Arctic. Panarctic began its exploration program with seismic work and then drilling in the Arctic Islands. By 1969 the Drake Point gas discovery was Canada's largest gas field. Over the next three years Panarctic discovered other large gas fields in the islands. These and later Arctic Islands discoveries contain established reserves of 17.5 trillion cubic feet of dry, sweet natural gas. Exploration moved offshore when Panarctic began drilling wells from "ice islands," platforms of thickened ice created in winter by pumping sea water onto the polar ice pack.

Oil has also been discovered on the islands at Bent Horn and Cape Allison, offshore at Cisco and Skate. In 1986, Panarctic company became a commercial oil producer on an experimental scale. This began with a single 100 000 barrel tanker load of oil from the Bent Horn oil field (discovered in 1974 at Bent Horn N-72, the first well drilled on Cameron Island). The company delivered its largest annual volume of oil, 300 000 barrels, to southern markets in 1988. The company now relies entirely on Arctic oil production for income, a notable landmark in frontier development.

Panarctic's ice island wells were not the first offshore wells in the Canadian north. In 1971, Aquitaine (later known as Canterra Energy, then taken over by Husky Oil) drilled a well in Hudson Bay from a barge-mounted rig. Although south of the Arctic Circle, the well was in a hostile frontier environment. A storm forced suspension of the well, and the ultimately unsuccessful exploration program languished for several years.

MACKENZIE DELTA AND BEAUFORT SEA

While Panarctic conducted early exploration in the Arctic Islands, preliminary investigations were underway north of Norman Wells in the Northwest Territories.

The Mackenzie delta was a focus of ground and air surveys as early as 1957, and geologists drew comparisons then to the Mississippi and Niger deltas, speculating that the Mackenzie could prove as prolific. For millions of years sediments had been pouring out of the mouth of the Mackenzie, creating tremendous banks of sand and shale – laminates of sedimentary rock warped into promising geological structures. Drilling began in the Mackenzie Delta-Tuktoyuktuk Peninsula in 1962.

In 1967 an event in Alaskan territory made it suddenly easier to raise money for petroleum exploration in the western Arctic. Arco Humble #1, a wildcat spudded at Prudhoe Bay, struck oil. The news did not instantly turn the languid investment situation around, but when another well struck oil several kilometres away, heads turned quickly. The Prudhoe Bay field proved to be the largest oil field in North America. For those exploring in nearby Canadian lands, it was evidence of the vast petroleum potential of the Arctic. As a result, drilling in the Mackenzie delta and the Beaufort Sea received a tremendous boost. Although the Mackenzie River delta did not compare to Prudhoe Bay, it did contain large gas fields. By 1977, its established gas reserves were 7 trillion cubic feet.

The level of activity in the delta dropped sharply in 1977. The proposed Arctic Gas pipeline died that year, and the federal government had no clearly defined energy policy for the frontiers, despite years of pressure from industry. Companies were unsure whether the eventual economics of production could justify further investment in the area. Finally, there had been no big oil discoveries. This was a critical drawback to exploration in that area, since easily marketable oil is the lifeblood of an industry which needs constant infusions of capital for reinvestment.

As in the Arctic Islands, onshore exploration eventually led to offshore drilling. However, trickier drilling faced those who wished to explore the lands under the Beaufort Sea. The petroleum industry gradually shifted its focus into the unpredictable waters of the Beaufort. To meet the challenges of winter cold and relatively deep water, drilling technologies in the Beaufort underwent a period of rapid evolution.

OFFSHORE DRILLING

The first offshore wells drilled in the Beaufort used artificial islands as drilling platforms. These artificial islands were a winter drilling system and only practical in shallow water. In the mid-1970s, the introduction of a fleet of reinforced drillships extended the drilling season to include the 90 to 120 ice-free days of summer. This also enabled the industry to drill in the deeper waters of the Beaufort Sea. By the mid-1980s, variations on artificial-island and drilling-vessel technologies had extended both the drilling season and the depth of water at which the industry could operate and had also reduced exploration costs.

The first well to test the Beaufort was not offshore, but was drilled on Richards Island in 1966. The move offshore came in 1972-73 when Imperial built two artificial islands for use in the winter drilling season. The company constructed the first of these, Immerk 13-48, from gravel dredged from the ocean floor. The island's sides were steep and eroded rapidly during the summer months. To control the erosion, the company used wire laid across the slopes and anchored, then topped off with World War II surplus anti-torpedo netting. The second island, Adgo F-28, used dredged silt. This proved stronger. Other artificial islands used other methods of reinforcement.[2]

In 1976, Canadian Marine Drilling Ltd., a subsidiary of Dome Petroleum, brought a small armada to the Beaufort. It included three reinforced drillships and a support fleet of four supply boats, work and supply barges and a tugboat. This equipment expanded the explorable regions in the Beaufort Sea. Drillships, however, had their limitations for Beaufort work. Icebreakers and other forms of ice management could generally conquer the difficulties of the melting icecap in the summer. But after freeze-up began, the growing icecap would push the drillship off location if it did not use icebreakers to keep the ice under control.

The major Beaufort explorers – Gulf Canada Resources Ltd., Dome Petroleum, Petro-Canada – experimented with new technologies and produced some of the most costly and specialized drilling systems in the world. Some were extensions of artificial island technologies as design engineers concentrated on ways to protect the island from erosion and impact. In shallow water, the standard became the "beach island." This island had long, gradually sloping sides against which the vengeance of weather and sea could spend themselves, gradually eroding the expendable islands.[3]

Although the artificial island was the best option for drilling in the Beaufort Sea, it was only practical in shallow water (no more than 25 metres deep). In addition, artificial islands required enormous volumes of material dredged from the sea bottom, and was therefore extremely expensive to build. To reduce costs, engineers began to look for better ways to construct these islands.

A deep-water innovation was made at Tarsuit, where Gulf constructed an island with a sand base or berm upon which four large concrete boxes or caissons were set in a square. Once in place, the company filled the caissons and the area they enclosed with sand. Tarsuit Island was an engineering success, requiring far less sand fill than other construction systems, and was also a geological success – a promising oil and gas discovery.

GULF'S DRILLING SYSTEMS

In a variety of forms, the caisson-retained island has since been used successfully at other Beaufort Sea locations. One variation was Gulf's 90-metre square floatable island, Molikpaq. The company would tow Molikpaq to a well location, then use sea water as ballast to lower it onto a foundation dredged out of the sea bed. Molikpaq's hollow core was filled with the material from dredging the sea, to provide stability. Following drilling, the operator refloated the unit, then towed it to the next location.

Another innovation, also developed by Gulf, is a hybrid between the drillship and the caisson-protected island. Kulluk is a conical vessel which provides island-like protection during deep-water drilling but has several advantages over other Beaufort drilling systems. As Kulluk is a vessel which is anchored during drilling, it does not need to settle on the sea floor before drilling can begin. Unlike artificial islands, it can drill in relatively deep water and unlike drillships, has a long drilling season.

The main disadvantage of the Kulluk system is that the stability provided by its circular design has made it a very "stiff" vessel. In its nautical sense, this refers to Kulluk's tendency to right itself quickly and even violently in rough seas. This can be quite hard on drilling equipment. However, the vessel's stability is a major advantage in field ice.

The Canadian Arctic as a Commercial Venture

Despite its high costs, Beaufort drilling continued into the mid-1980s, resulting in important discoveries. However, when the Canadian Arctic will become a commercial production area for petroleum is a matter of much dispute. The industry has made several notable efforts to demonstrate the viability of production from the Arctic, but only one has met with modest success.

In the 1970s and early 1980s, there were three major pipeline proposals to deliver natural gas from the Beaufort to market. These included proposals for pipelines down the Mackenzie Valley from Canadian Arctic Gas, a competing proposal (the Maple Leaf line) from Foothills Pipelines Ltd., and a proposed Polar Gas Project, which would transport gas from both the Arctic Islands and the Mackenzie Delta on the shores of the Beaufort Sea. In addition, a proposed Arctic Pilot Project would have liquefied natural gas from the Arctic Islands, and shipped it to market by ocean-going tanker. In time, each of these projects was shelved. Deteriorating oil and gas economics made them first marginal, then clearly uneconomic.

But three events occurred in the 1980s which were of more than passing significance to Canada's northern frontiers:

1. The large-scale, $530 million development of the Norman Wells oil field, complete with a pipeline to Canada's southerly pipeline system. Planning began in 1978, but approvals and construction kept the field from full production until 1985. After serving a small regional market for 65 years, Norman Wells became an important contributor to Canada's oil supply, and rapidly reached peak productive capacity of 22 000 barrels per day.
2. In 1985 Panarctic delivered a shipment of crude oil from the Bent Horn oil field in the Arctic Islands to Montreal. The 100 000 barrels of oil travelled by a reinforced vessel to Little Cornwallis Island, then by standard tanker to Montreal. By itself, the field will not cover total High Arctic exploration costs. However, as a stand alone project, Bent Horn is a commercial oil field. It covers all of Panarctic's overheads, operating costs, transportation and field developments, and has done so since 1986. Bent Horn has demonstrated that producing and delivering oil from the Arctic Islands is more than a dream.
3. In September 1986 Gulf delivered the largest purchase of oil from the Arctic, 320 000 barrels of Beaufort Sea oil to Japan. The oil was produced during extended production tests of the Amauligak well. These tests demonstrated that Amauligak was a prolific reservoir, but also caused Gulf to reevaluate its reserves estimates downward. Shortly afterward, Gulf mothballed its Beaufort Sea operations.

None of the Beaufort explorers have formally proposed shipping oil to Prudhoe Bay where the existing Valdez pipeline could transport production to warm-water tankers. Although perhaps not politically palatable, this could be the least expensive way to begin development of the Beaufort Sea.

OFFSHORE DRILLING RIGS

The first offshore well in the world was drilled from a wooden pier on the California coast in 1887. Later wells moved offshore completely; wooden derricks stood atop wooden platforms in shallow water. After World War II, wooden platforms became obsolete as steel platforms became the standard. The fixed drilling platforms used today are larger and more complicated still. These stronger structures helped to advance the offshore frontier farther out to sea.

Mobile drilling platforms began to appear. The first of these were submersible platforms floated into place and sunk to rest on the bottom. The next generation consisted of "jack up" rigs floated into place, then jacked up on steel legs or frames resting on the ocean floor. Later came semi-submersible platforms which could drill in harsh environments, with perhaps 25 metres of the vessel's legs submerged. Then came drillships with rigs mounted on deck. These progressive innovations allowed the industry to move into ever-deeper water.

Drillships and semi-submersibles are the only vessels capable of drilling in water of more than 300 metres deep. To enhance this capability, the vessels use dynamic positioning – a high-tech system which stabilizes the rig on one spot without the need for anchors.

THE EAST COAST OFFSHORE

The site of Canada's first saltwater offshore well was 13 kilometres off the shores of Prince Edward Island. Spudded in 1943, the Hillsborough #1 well was drilled by the Island Development Company, a subsidiary of what is today Mobil Oil Canada. The company used a drilling island constructed in 8 metres of water, made of wood and some 7 200 tons of rock and concrete. The well was drilled down 4 479 metres at a cost of $1.25 million, an extremely expensive well in that era. Hillsborough, which had been part of the Allied war effort, was declared dry and abandoned in September 1945.[4]

In the post-war period, Mobil has continued to be an active player in the east coast offshore. The company first applied for offshore permits in 1958, and acquired rights to 445 000 hectares off Canada's east coast, the area including and surrounding Nova Scotia's Sable Island. The offshore portion of this area reached water depths of 30 metres. This signalled the beginning of unprecedented expansion of frontier drilling lands. Leasing activity spread north from Sable Island to Labrador. By 1975 various exploration permits on the Atlantic offshore covered 1.3 million square kilometres.

The geological theories of "continental drift" and "sea floor spreading" suggest that Europe, Africa and the Americas were originally one land mass and helps to explain why some of the same sediments are found on distant continental shelves. Upheavals from the earth's core caused the original land masses to break apart and form a basin which became an inland sea. Over tens of millions of years, after successive eruptions from the earth's core, the sea floor spread apart, pushing the shorelines of that early inland sea farther and farther apart. Those ancient seabeds formed the rocks which, combined with sediments from rivers and shoreline erosion, make up the continental shelves and their adjacent slopes.

The Canadian sections of the ancient sea are the Scotian Shelf, the Grand Banks off Newfoundland and the Labrador Shelf. They also include adjacent deep-water areas of the continental shelf. In Europe, sedimentary basins are found in the Irish, Baltic and North Seas, and along the coast of northern Norway. The Grand Banks and (especially) the North Sea have been prolific hunting grounds for the petroleum industry.

The Scotian Shelf

In 1967 Shell drilled the first well off Nova Scotia, the Sable Island C-67. Located on desolate, sandy Sable Island (best known for its herd of wild horses), the well bottomed in gas-bearing Cretaceous rocks. Drilling stopped because the technology did not exist to handle the pressure encountered. Today, the blow-out preventers required for offshore work in eastern Canada can handle working pressures of 15 000 pounds per square inch.

Shell's experience with this well foreshadowed two future developments on the Scotian Shelf: major discoveries off the Nova Scotia shore would be gas reservoirs and would involve high-pressure natural gas.

West Venture started as a surface blowout and was swiftly shut in. But the well then blew out underground. High-pressure natural gas burst through the well's casing, and began rushing from a deep zone into a shallow one. In oil industry parlance, the blowout "charged" the shallower geological zone, drastically increasing reservoir pressure. The cost of bringing this one well under control was a phenomenal $200 million.

The industry made other modest oil and gas discoveries in its early years off Nova Scotia – for example, Shell's Onandaga E-84 gas well, drilled to a depth of 3 988 metres in 1969 and Mobil's 1973 D-42 Cohasset well on the western rim of the Sable sub-basin.

At Logan Canyon, Mobil's drill bit found almost 50 cubic metres of net oil pay in 11 zones of Cretaceous lower sands. However, a follow-up well five years later found only water-bearing sands, and the company suspended work on the field. Mobil moved to other Scotian Shelf locations, discovering the promising Venture gas field in 1979.

Located on a seismic prospect which had been recognized some years earlier, Mobil waited to drill the West Venture N-91 probe until the early eighties because the structure was deep and could contain high-pressure zones like those which had halted drilling at Sable Island in the previous decade. The West Venture discovery well cost $40 million, at that time a startling price for a single well. When the well blew wild, it took eight months to bring it under control. By contrast, another discovery well drilled around that time, Shell's Uniacke G-72, blew wild and was controlled within 10 days.

Ironically, the first commercial offshore well, Mobil's 1973 Cohasset discovery, appeared relatively inconsequential when found. But toward the end of the 1980s, a combination of exploration successes and innovative thinking led to development of a field which most of the industry had seen as uneconomical.

In December 1985, Petro-Canada spudded the Cohasset A-52 step-out well to explore the Cohasset structure southwest of Mobil's 1973 discovery well. Unlike the disappointing 1978 step-out, that hole tested oil at a combined rate of 29 000 barrels per day from six zones.

Following up on the positive results of the A-52, Shell drilled a discovery well at Panuke, eight kilometres southwest of Cohasset. The Shell Panuke B-90 wildcat encountered a relatively thin zone that tested light oil at a rate of 6 100 barrels per day. The following year, Petro-Canada drilled the F-99 delineation well at Panuke. That well tested oil at 4 100 barrels a day for six days.

While the Cohasset and Panuke discoveries were marginal by themselves, a consulting firm hired by Crown corporation Nova Scotia Resources Limited seized on the idea of joining them together. By forming a joint venture with British-based LASMO plc., which formed a Nova Scotian affiliate to operate the field, NSRL was able to make the project a financial and technical success.

Despite the geological successes in the Scotian Shelf, the economic prospects for offshore development crashed in early 1986 as oil prices dropped from $30 to less than $15 per barrel. Despite the crisis precipitated by this drop in prices, several factors made the Cohasset project economical. Geologically, it had excellent reservoirs. Physically and chemically, the oil had exceptional fluid characteristics and was highly desirable to refineries. In addition, the field was in relatively shallow water. This made it possible to develop using relatively small wellhead platforms and to use a converted jack-up rig for the production facility. Finally, LASMO leased such major equipment as a jack-up rig (which it converted into a production platform) and tankers. This kept development costs relatively low.

After field development began, LASMO continued exploratory drilling, locating the Balmoral field four kilometres from Cohasset. In combination with results from Panuke development drilling and Cohasset seismic reevaluation, this discovery increased estimated reserves for the three fields to 49 million barrels. The field went on production on June 6, 1992.

The Labrador Coast and the Grand Banks

The bitter-cold Labrador Shelf of Newfoundland and Labrador was another prospective realm for exploration during the early period of eastern offshore exploration. The Labrador Shelf was first drilled in 1971 using drillships positioned in the Labrador Sea. Icebergs from the glaciers of

Greenland and Labrador soon earned this stretch of water the nickname "Iceberg Alley." Icebergs drifting toward drilling equipment posed a unique hazard for the industry in that forbidding environment. Labrador drillers handled the problem by "lassoing" the bergs, cowboy-style, with nylon ropes and steel hawsers and towing them out of the way.

In the end, however, worsening exploration economics and poor drilling results dampened the industry's enthusiasm for the area. Drilling stopped in the early 1980s. It continued, however, in the more southerly waters off Newfoundland.

The most promising exploration off Canada's east coast has taken place on the Grand Banks, particularly the Avalon and Jeanne d'Arc basins. Exploration began in the area in 1966 and except for one oil show in 1973, the first 40 wells on the Grand Banks were dry. Then, in 1979, came the Hibernia oil strike, which appeared to change the fortunes of the area. Although not large enough to be commercial at the time of discovery, the next nine wildcats were important oil discoveries. And two discoveries in the mid-1980s, Terra Nova and Whiterose, may prove to be more inexpensively producible than Hibernia.

Chevron earned a commercial interest in the Hibernia field by spudding the discovery well in 1978. The field is 315 kilometres east-southeast of St. John's, and water depth is about 80 metres. Between 1980 and 1984, Mobil drilled nine delineation wells in the field at a cost of $465 million. Eight of those wells were successful, enabling the industry to establish that the field has recoverable reserves between 525 and 650 million barrels.

But bringing the field on production would still be a long time coming. It involved settling a jurisdictional dispute between Newfoundland and Canada over ownership of offshore minerals[5] and other issues. Lengthy fiscal negotiations began in 1985, shortly after Mobil submitted a development plan to the two governments. Not until 1988 did the two governments reach agreement on the development with Mobil, Petro-Canada, Chevron and Gulf, the companies with interests in the Hibernia field.

By the terms of this agreement, the federal government would provide $1 billion in grants, $1.66 billion in loan guarantees and other assistance to the $5.2 billion development project. These concessions were necessary because government insistence on a huge, expensive concrete production platform (the Gravity Based System) had made the field uneconomical.

While a floating platform like those used in the North Sea would be far less expensive, construction of a Gravity Based System is labor-intensive and also arguably safer for the iceberg-prone Grand Banks area. For government, the project is appealing from a regional development standpoint, since Newfoundland has chronically high unemployment. Whether profitable to the developers or not, Hibernia will generate revenue to stimulate the economy of Canada's poorest province.

Scheduled to begin producing 110 000 barrels of oil per day in 1997, the project met with a crisis in early February, 1992. Gulf gave notice that it was withdrawing from the project. The reasons Gulf cited were "constrained cash flow" (not have enough cash at its discretion to justify investing in a marginal project), deteriorated business conditions since industry and government had signed the Hibernia agreement and a six year wait until the project would begin to contribute to cash flow. Finally, Gulf said it had other attractive business opportunities and did not want to commit such a large portion of resources to a single project.

In accordance with the Hibernia agreement, Gulf continued paying its share of the bills while the remaining partners and the provincial and federal governments scrambled to find a company to take Gulf's place. In the end they were only partial successful.

In an eleventh-hour deal almost a year after Gulf's announcement, Arkansas-based Murphy Oil Corp. agreed to take a 6.5 percent interest in the project. The federal government agreed to take 8.5 percent, for an additional commitment of $290 million – $10 million less than the cost to government to wind down the project.

THE WEST COAST

A sedimentary basin also exists off the west coast of Canada, where some exploratory drilling has taken place. From 1967 to 1969, Shell drilled 14 deep dry holes, some west of Vancouver, others in Hecate Strait beside the Queen Charlotte Islands. Exploration off the west coast stopped in 1971 when the federal and British Columbia governments agreed to a moratorium on exploration pending the results of studies into the environmental impact of drilling. In 1986 a government-appointed commission recommended an end to the moratorium.

The province had still not acted by 1989, however, when an American barge spilled oil off the British Columbia coast. A few months later came the disastrous Exxon Valdez oil spill off Alaska. Although neither of these spills was related to crude oil exploration or production, they made it politically impossible for governments to lift the moratorium.

Although not truly part of Canada's petroleum frontiers, there has also been some activity in the adjacent Fraser Delta, on B.C.'s lower mainland. This area had gas shows as early as 1904 when the Steveston Land and Oil Company drilled there in search of oil. In the early 1920s, industry pioneer Frank McMahon had exactly the same experience. But because gas was considered useless in those days, the area remained largely unexplored and in the 1970s the British Columbian government placed a moratorium on exploration in the area, for environmental reasons.

However, in 1990 the provincial government struck a Commission of Inquiry to investigate the environmental and economic impacts of drilling in the sedimentary delta. Less than a year later, the commission reported that there was virtually no environmental risk, and that oil or gas finds would contribute to the local economy. The government lifted the moratorium and gas exploration began.

THE COSTS AND THE BENEFITS

In exploring its frontiers, Canada has drilled some of the world's deepest offshore wells, built new artificial island and mobile drilling systems and created networks capable of providing instant communication between head office and remote well sites. Canada has developed the world's most sophisticated knowledge of methods to deal with ice while exploring for oil in the north.

These initiatives and others have enabled the Canadian petroleum industry to develop unrivalled expertise. But offshore drilling has become expensive, as illustrated in the following table:

FRONTIER DRILLING COSTS* 1960 - 1990

Period	Yukon and Northwest Territories			East Coast Offshore		
	Exploration Spending	Wells	Cost per well	Exploration Spending	Wells	Cost per well
1961-65	$ 86	65	$ 1.3	–	–	–
1966-70	$ 270	223	$ 1.2	$ 69.3	16	$ 4.3
1971-75	$1201	312	$ 3.8	$ 345.7	95	$ 3.6
1976-80	$1901	85	$22.4	$ 779.8	32	$24.4
1981-85	$4491	98	$45.8	$5052.3	80	$63.2
1986-90	$1394	68	$20.5	$1369.6	31	$44.2

*All dollar figures in millions
Source: Canadian Petroleum Association, 1992

This increased expense left the industry's frontier operations highly vulnerable to the 1986 decline in oil prices. Costly frontier drilling, which had found reserves that were marginal at best in the lower-price environment, was the first casualty of the industry-wide crisis. A precipitous decline in frontier activity was well underway by mid-1986, and drilling was almost at a standstill by year end. However, activity did not cease and costs (inflated during the 1981-85 period by federal Petroleum Incentives Payments) quickly declined.

As exploration in the frontiers wound down, the industry began to prove that its expertise was saleable around the world by becoming active in what was then the Soviet Union.

In 1987, the Soviet Union created joint venture legislation which permitted foreign industry to operate in that country for the first time. The very first western company to ink a petroleum-related deal with the Soviets was Canadian Fracmaster, a Calgary-based service company. The Fracmaster joint venture involved servicing 4 500 wells in western Siberia, thus increasing well productivity. The first Canadian oil company to begin developing Soviet oil fields was Gulf Canada. In 1989 Gulf signed a deal to increase production from two light sour oil fields north of the Arctic Circle, on the eastern edge of Europe.

With the 1991 collapse of the Soviet Union into its constituent republics, more players from the Canadian petroleum industry began looking seriously at ventures in the republics for opportunities to grow. And much of the advantage Canadians have in negotiating these deals comes from its experience in its own frontiers – first in the near north, and later in the far north and the offshore.

Chapter 10

From Crude to Refined

Imperial Oil Refinery, Whitehorse, NWT, 1946.
(Provincial Archives of Alberta, H. Pollard Collection/P1356)

The oil industry is described as having "upstream" and "downstream" sectors, terms that evoke images of oil flowing to the end user from wells through intricate networks of pipes, refining vessels and other paraphernalia. The "upstream" end looks for and produces oil and gas. The "downstream" refines and markets oil products and natural gas.

Depending on demand, refining technology and the kind of oil used, the downstream refiner makes many products from every barrel of oil. These include gasoline, jet fuel, home heating oil, lubricating oils, greases, asphalt and roofing tar.

People have probably been experimenting with raw oil to fit it more closely to their needs since paleolithic times. The first writer to mention a simple form of refining was Herodotus who in 450 B.C. referred to a well in Kir Ab in Susiana (Iran) which:

> *... produces three substances, for asphalt, salt and oil are drawn from it in the following manner: it is pumped up by means of a sweep, and instead of a bucket, half a wine skin is attached to it. Having dipped down with this, a man draws it up and then pours the contents into a reservoir, and, being poured from this into another, it assumes these different forms: the asphalt and the salt immediately become solid, but the oil they collect.*[1]

Modern refining owes part of its heritage to the nineteenth century depletion of whales. Whale oil had long been serving the lubricating needs of burgeoning industry and the demand for home lighting fuel. Whale populations declined under the pressure of hunting. And as the whalers drove their prey to more distant ocean frontiers, short supplies coupled with growing demand pushed whale oil prices high. This led scientific dabblers to search for less expensive sources of lamp fuel and lubricating oil. The idea of creating a naphtha distilled from coal tar appeared as early as the 1780s. In 1820, a German distilled a wax he called paraffin from coal tar and wood tar.

Canada made an impressive debut in this area when Nova Scotia's Dr. Abraham Gesner made an illuminating oil from coal in 1846. He later created other oil products from both Trinidad bitumen and New Brunswick albertite. He called his product kerosene.

In 1853 Gesner went to the United States. There, his patents launched the North American Kerosene Gas and Lighting Company a year later. This company's Long Island refinery produced kerosene from coal, hence the name coal oil, often still used. Gesner, whose work laid the base for early refining, returned to Halifax a wealthy man.[2]

James Young of Scotland distilled a similar fuel product (which he called "paraffine") out of crude oil in 1848 and received a patent just before Gesner. Young's petroleum source was not reliable, however, and so he began using coal as his feedstock, receiving British patents on his paraffin or coal oil process in 1850. J.T. Henry wrote in his 1873 *Early and Later History of Petroleum*:

> *Uniting himself with capitalists, Mr. Young promptly began the manufacture of paraffine oil on a large scale. No lamp had yet been invented in which it would burn without a most offensive smoke, and while the heavier of this manufacture was oil for lubricating machinery, the lightest was reduced to paraffine wax, manufactured into candles, and sold as spermaceti, to which it bears a striking resemblance. It is a product obtained by destructive distillation of the oil – that is, one substance is destroyed before the other is produced.*[3]

In 1855 a group of businessmen asked Yale professor Benjamin Silliman to prepare a report on the chemical properties of Pennsylvania rock oil found in seeps. Silliman submitted the oil to fractional distillation and produced eight products of different density and boiling point. None of these distillates behaved very well in any lamp available. Nonetheless, Silliman's work was a key factor in drilling the first American oil well four years later.

The discoveries of free oil later in the decade – in the Enniskillen gum beds of Ontario in 1958 and in Pennsylvania the following year – led to the construction of numerous primitive refineries. Five years after Williams drilled the first North American well, there were 60 refineries in Pennsylvania, 30 in Ontario.

EARLY REFINING

Refining in Canada, as in the United States, began slightly before the discovery of free oil by drilling. Dr. Gesner was making his kerosene from tar and bituminous oil in the 1840s. And a simple version of refining was taking place in the Enniskillen gum beds near Sarnia, Ontario, where refiners boiled the petroliferous gum in cast-iron kettle stills to rid the asphalt of lighter petroleum fractions. In fact, James Miller Williams, early oil pioneer, was developing gooey oil from these marshlands for just this kind of oil processing when his water well turned into an oil well in 1858.

Canada's first petroleum refinery was a simple cast iron affair in the village of Oil Springs. By 1863, there were 10 refineries in Oil Springs, 5 in Toronto and perhaps another 15 scattered throughout Ontario.[4] These refineries processed crude oil that made its perilous, gummy way from the field in wooden barrels on horse-drawn wagons.

Canadian refiners were quick to seek export markets. However, the foul smell of their oil products – an odor which resulted from high levels of sulphur in southwestern Ontario crude – hampered their efforts. The thriving Pennsylvania industry produced oil that contained far less sulphur than its Canadian counterpart.

The newly formed Dominion of Canada introduced an import tariff in 1867 to protect its refiners from having to compete at home with cleaner, better-smelling products from south of the border.[5] J.T. Henry reported that an English technique known as Allen's Process helped deodorize Canadian oil.

What "Allen's Process" may have been is a matter of conjecture. What is clear, however, is that the Canadian industry developed the litharge process, an oil-sweetening process that uses lead oxides. William Spencer of Woodstock, Ontario developed this process in association with petroleum pioneer James Miller Williams. The litharge process was widespread in the Canadian industry for the final decades of the nineteenth century. The litharge process (known as "doctor sweetening") became an important part of the refining process in the twentieth century but there has been little recognition that it began in Canada.[6]

Canada had begun to export refined products by 1868. And by 1870, Canada exported about 60 percent of domestic illuminating oils and lubricants to Europe.[7] But those exports dwindled over the next several years because of the poor burning qualities of Canadian oil products, which could not compete with better quality American products. Although the smell was mostly gone, Canadian refined oil still contained tar and sulphur compounds, and therefore burned with an undesirable, oily smoke.[8]

Lacking major discoveries, the Canadian industry remained small. A few important refining companies did emerge, however. In 1880, 16 Canadian refiners banded together to form the Imperial Oil Company. In 1904, some smaller Canadian independents, producers and refiners formed the Canadian Oil Company Limited. And in 1906, Canadian interests formed the British American Oil Company.[9]

In time, work begun by Imperial Oil completely resolved the problem of "sour" oil – oil with sulphur present. Herman Frasch, an engineer Imperial hired to work on the sulphur problem, discovered a way to remove sulphur from oil almost entirely. Ironically, as Frasch resolved the problem, American wildcatters brought in the first large sour oil field in the United States, on the Ohio/Indiana border. So Frasch's discovery was timely, for he almost immediately began working for Standard Oil to apply the process to production from that field.[10]

THE AUTOMOBILE: CRACKING AND OCTANE RATING

In the early years of the twentieth century, the United States was the undisputed pioneer in the refining business. American dominance came from vast oil production, business acumen and increasing automobile sales in the United States. Automobiles radically changed the refining industry. In Canada, Europe and everywhere else, refining technology lagged far behind.

Until automobiles created the demand for automotive fuels, refiners had been almost exclusively in the business of making lubricants and lighting fuels. But as the use of automobiles grew, the internal combustion engine found important new applications during the First World War. Accordingly, refiners shifted their focus to transportation fuels. They became increasingly concerned about getting the maximum gasoline possible from their crude oil feedstock.

The first efforts to produce gasoline consisted of adding condensing towers to the basic refinery stills. Then, in 1912, the Standard Oil Company used heat and pressure to "crack" the larger and heavier hydrocarbon molecules. In this way, the company could get a higher percentage of the lighter (and more valuable) distillates from the oil.

At least in theory, thermal cracking was not new in 1912. There is a popular legend that an American stillman accidentally discovered cracking when he left oil distilling under pressure for too long in 1861. And the 1855 report on Pennsylvania rock oil by Dr. Silliman suggests that the oil would crack into a lighter product under sustained heat and pressure. However, commercial development of hydrocarbon cracking went unexploited until automobile use became widespread.

During the Second World War, many companies strove to develop cracking processes of their own to avoid licensing the Standard Oil patent. Cracking was widespread by the 1920s, and the gasoline yield from a barrel of oil increased greatly as a result.

The development of more powerful automobile engines created greater demand for better gasoline, gasoline that would prevent engine "knock." The industry term for a gasoline's anti-knock properties is its octane rating. In developing the cracking process, the industry learned that cracked gasoline had a higher octane rating than did gasoline produced through normal distillation. It also found that adding tetraethyl lead improved the octane rating of gasoline.

Another route to high octane was polymerization, the combination of similar light hydrocarbon molecules to form heavier hydrocarbons. The development of these processes resulted from the growing demand for high-octane fuel. Besides producing more gasoline, cracking produced light, unstable hydrocarbons. Polymerization could combine these compounds into heavier, high-octane fuel components.

In the late 1930s, the industry found ways to react the off-gases produced during cracking to form high-octane fuels. One of these was alkylation, the reaction of iso-butylene with n-butylene. In 1937, Frenchman Eugene Houdry developed a commercial method of catalytic cracking, again revolutionizing the oil refining industry. Catalytic cracking of oil in the presence of certain catalysts both improved gasoline's octane rating and lowered its sulphur content.[11]

These developments were widespread when the Second World War began. Due to the critical need for 100-octane gasoline for warplanes, North American 100-octane refining capacity doubled every 12 months during the war.[12]

CANADIAN REFINING BETWEEN THE WARS [13]

After the First World War, demand for gasoline soared. Canadian oil imports and refined product imports rose to keep pace. Output from Ontario, New Brunswick and Alberta oil fields was too small to meet the demands of Canadian refineries, so the fuel for Canadian cars and trucks derived mainly from imported oil. Canadian refineries relied heavily on American oil, and marketers also imported gasoline and other products from the United States. Canadian reliance on the United States continued until the mid-1900s.

Some domestic crude oil fed refineries in southern Ontario, and a few tiny refineries in western Canada processed crude from small local oil fields. In 1923, an Imperial facility in Calgary began refining Turner Valley condensate.

Imperial Oil, which had become part of Standard Oil in 1898, was the dominant force in early refining and marketing in Canada. In 1927, Imperial owned 90 percent of Canada's refining capacity. Its oldest and largest refinery, located at Sarnia, Ontario, refined crude which came from the American Midwest and southern Ontario. A Regina refinery relied exclusively on American imports. Imperial operated three tidewater refineries – at Halifax, Montreal and Ioco near Vancouver, each refining imported oil.

Not surprisingly, Imperial achieved many Canadian refining firsts: the first battery of condensation stills at Montreal in 1917, the first continuous thermal cracking stills at Calgary in 1923, the introduction of tetraethyl lead in 1926, an early catalytic cracker at Sarnia in 1940, and the first alkylation plant at Calgary in 1944.[14] The company actually shared this last distinction with Shell's Montreal refinery, which also began alkylation in 1944.

When the Canadian tariff on gasoline imports dropped from 0.5 cents to 0.2 cents per litre in 1922, several Canadian companies started importing American refinery products and selling them in Canada. Some of these companies eventually became refiners themselves. McColl Brothers Limited had been importing gasoline from the United States since 1916, and entered the refinery business in 1925. Other Canadian refiner-marketers struggling to expand market share during the 1920s were the British American Oil Company (later Gulf Canada) and the Canadian Oil Company. Frontenac Oil Refineries originated as a refiner-marketer in 1925. In 1927, McColl Brothers and Frontenac merged to form McColl-Frontenac. During the 1930s McColl-Frontenac associated with the Texas Company and eventually became Texaco Canada.[15]

In the early 1930s, the government of Canada again raised the tariff on imported gasoline. The result was a quick boost in the number and capacity of Canadian refineries. By 1940, Imperial's share of refining capacity had declined to slightly more than half.

In western Canada, refining of domestic crude oil inched ahead during the 1920s and 1930s. In addition to Imperial's Calgary facility, Wainwright Producers and Refiners Ltd. began operating a small heavy oil refinery. In 1935, Imperial revived its efforts to refine Norman Wells oil, operating a 500 barrel-per-day refinery at Norman Wells to serve mining ventures in the north.

The situation changed in 1936 when the industry discovered the oil column at Turner Valley. Increasing oil production sustained new refineries in Calgary, as well as the Gas and Oil Products refinery in the Turner Valley field at Hartell. For a few months during the Second World War, the Canol pipeline transported Norman Wells oil to a refinery at Whitehorse, Yukon.

The technological development of the Canadian refining industry followed a pattern similar to that in the United States: thermal cracking and reforming to achieve high yields of high octane gasoline, and isomerization and alkylation to use natural gas by-products and refinery off-gases in high octane blends. During the Second World War, Imperial's Calgary refinery added an alkylation unit to supply the high demand occasioned by the Commonwealth Air Training project in western Canada.

AFTER LEDUC (POST-WORLD WAR II)

The Leduc oil strike in Alberta in 1947 and other later finds turned Edmonton into western Canada's largest refining centre. Partly because of post-war steel shortages and partly because a perfectly good refinery was available for a song, the first Edmonton refinery after Leduc was the Whitehorse refinery that had processed oil from Norman Wells during the last few months of the war. Imperial dismantled the refinery and reconstructed it in Edmonton in 1948, adding 10 000 barrels of daily capacity in 1949. British American Oil (later Gulf) was the next to build a refinery in Edmonton, completing it in 1951. The same year, McColl-Frontenac (later Texaco) completed an Edmonton refinery, its first western Canadian operation.

With the completion of crude oil pipelines from Edmonton to central Canada and to the west coast in 1951 and 1953 respectively, Alberta oil began to feed refineries in every province from Ontario west. And Canadian exports began supplying refiners in the United States.

In 1946, Canada was producing less than 10 percent of the 220 000 barrels of oil Canadians consumed each day.[16] Ten years later, domestic oil supplied 55 percent of the much-larger Canadian market. National refining capacity had tripled during the same decade.[17] And gasoline demand and refinery capacity continued to grow. Facilities became increasingly sophisticated while products improved and became more varied in their use.

The 1960s saw five new refineries completed. One of these deserves special mention. Irving Oil's refinery in St. John, New Brunswick, began operations in 1960. Today it is by far the largest refinery in Canada, with capacity of nearly 250 000 barrels per day.

Another 1960s expansion project is noteworthy. Imperial Oil expanded operations at its Sarnia refinery, one of the largest in Canada, by 16 000 barrels, to 110 000 barrels per day. That development was an exercise in building upon history, since the refinery has served Canadian needs since 1897.

The 1970s were the decade of the most dramatic growth in refining. Five new refineries began operations – each one of Canada's largest. There were also countless refinery expansions. But so much construction led to overcapacity and in the 1980s Canadian refining capacity began to decline in response to the previous decade's overbuilding.

In 1985, Shell Canada developed the first refinery ever designed solely for the processing of synthetic oil from the Athabasca oil sands. Located at Scotford, northeast of Edmonton, this 60 000 barrel-per-day facility has been a notable success. Although paying a premium for its feedstock, the refinery can squeeze more light products from synthetic feedstock than from conventional oil. The smaller waste (no asphalt or bunker fuels from the refining process) means that the refinery gets a better return from every barrel of oil refined.

COMPETITION AND CONTRACTION

The most important reasons for the rapid growth in refining capacity in the post-war years were the general prosperity that followed the Second World War and increasing demand for automobile and other transportation fuels.

Until the early 1970s, sales of refined products increased steadily around the world. However, for most industrial nations that trend came to a halt in 1973 when the first OPEC engineered oil price shock raised the price of oil. The severity of the price increases led to deteriorating demand for refined products in most consuming countries. For example, oil consumption peaked in 1973 in both Japan and western Europe, then began more than a decade of steady decline. Two decades later, these countries are still consuming far less oil than they did in the early 1970s.

However, federal oil pricing policy muted the effect of the first oil price shock on Canadians. Although policy-makers believed the world had entered an extended and chronic period of crude oil shortages, the federal government developed policies which effectively subsidized consumption. Thus, demand continued to increase.

Canadian consumption finally began to decline in 1980. For the first time on record, Canadian refineries processed less oil than the year before. Spurred by a second oil price shock, this shift came for several reasons:
1. The gradual increase in the fuel efficiency of Canada's fleet of automobiles began to have a cumulative affect.
2. Motorists were reluctant to drive as much, since gasoline prices were climbing steeply.
3. People applied conservation technologies like home insulation more widely.
4. Canadians began substituting fuels like natural gas and electricity for heating oil.

In 1985, new Canadian energy policies allowed oil prices to rise to world levels, giving product prices an additional nudge upwards. In addition, the federal government dropped restrictions on the export of refined products. Thus, while Canadian refiners found themselves facing declining domestic demand, they also found the limitations on exports removed. Exports did not greatly increase, however, since demand was also declining in the United States, Canada's largest products market.

Oil prices crashed in 1986 (the third oil price shock), but the consumer received only temporary relief. Federal and provincial gasoline taxes had been rising steadily throughout the decade. After oil prices dropped they rose with renewed vigor. For example, average federal taxes on gasoline doubled in the first half of the 1980s, to just more than 5 cents per litre in 1984. They more than doubled again by 1989, to 11.3 cents per litre. Average provincial taxes also doubled during the decade.[18] Thus, governments quickly took up most of the slack in gasoline prices after crude oil prices plummeted. Gasoline's pump price declined only slightly. And because prices stayed high, demand continued to decline.

In 1979, Canadian refineries processed just under 2 million barrels of oil per day, and Canadians consumed about 1.9 million barrels of oil products. By 1991, refinery runs were down to 1.5 million barrels, and consumption was less than 1.4 million barrels. So successful was the shift away from heating oil in those years that consumption dropped by nearly half. Today that product accounts for only about 7 percent of refinery output (down from 11 percent.)

Declining demand did more than bring refinery construction to a halt. It also speeded up the contraction of refining capacity which began in the early 1980s. The years since 1973 have seen the closure of many small and technologically outdated facilities. These plants could not compete technologically or economically with large new refineries, and environmental considerations added to their woes. The following table, which summarizes new refineries and closures since 1970, illustrates the shift toward larger, more efficient refineries.

REFINERY OPENINGS AND CLOSURES IN CANADA

Years	New Refineries	Average Capacity (barrels per day)	Refinery Closures	Average Capacity (barrels per day)
1970-74	3	115 000	3	7 850
1975-79	2	135 000	7	32 800
1980-84	2	44 300	10	37 900
1985-91	0	–	2	44 750

Source: Canadian Petroleum Association, 1992

While Canada's first refineries were tiny operations, sometimes as small as 50 barrels per day, today's are large and efficient. They are hundreds or thousands of times larger than pioneer plants constructed early in this century.

When the scales weighing product demand and refining capacity finally find a balance, the large and technically sophisticated refineries in the following table will certainly survive to supply Canadian markets.

CANADA'S LARGEST REFINERIES

Owner	Location	Capacity (barrels per day)	Year Commissioned
1. Irving Oil	Saint John, NB	250 000	1960
2. Imperial Oil	Strathcona, AB	164 900	1975
3. Imperial Oil	Sarnia, Ont.	121 500	1897
4. Ultramar	St. Romauld, PQ	120 000	1971
5. Shell Canada	Montreal, PQ	120 000	1933
6. Petro-Canada	Edmonton, AB	115 500	1971
7. Newfoundland Refining	Come-by-Chance, NFLD	115 500	1973
8. Imperial Oil	Nanticoke, Ont.	106 300	1978
9. Petro-Canada	Montreal, PQ	87 500	1955
10. Sunoco	Sarnia, Ont.	83 100	1953

Source: Canadian Petroleum Association, 1992

GASOLINE MARKETING

The first car probably rolled down the dusty road in Canada in 1898. An item from the *Edmonton Bulletin* of March 2, 1906 illustrates the inefficiency of those early vehicles. The occasion was the first recorded automobile trip from Edmonton to Calgary – on today's roads, a distance of about 300 kilometres. The party left Edmonton on Saturday morning at ten and arrived in Calgary at seven on Sunday evening, staying in Red Deer over night. From Lacombe to Red Deer, a distance of 32 kilometres, took 34 minutes by car, so fast that it deserved special mention from the *Bulletin's* reporter. During the trip, 80 litres of gasoline were used and four litres of lubricating oil.[19]

Motorists in those days bought gasoline from general stores. The gasoline service station, dispensing both fuel and repairs, made its American debut in 1907. Imperial opened the first Canadian gas station in 1908.

Because the quality of gasoline differed markedly in the early years, brand names became extremely important. They were the consumer's only guarantee of quality. Refiners operated most early service stations themselves, but the drop in the gasoline import tariff in 1922 enabled a few non-refiners to enter the business. However, Imperial – Canada's largest oil producer and refiner – has also always been its largest retail marketer. From the beginning, Imperial has been the giant among Canadian oil companies.

By 1934, two-thirds of Canada's 2 300 service stations were in the hands of the five major oil refiners: Imperial, Shell, British-American, McColl-Frontenac and Canadian Oil. Irving and Supertest were prominent in the Maritimes.

After the Second World War, the number of automobiles in Canada increased dramatically, as did the size of cars. The number of vehicles rose by 94 percent during the 1950s and increased by another 60 percent during the following decade. Reflecting cheap and plentiful oil and general prosperity, the cars of those years were the infamous "gas guzzlers."

The most popular of the rapidly proliferating service stations was the full service station. This early convenience store offered tires, batteries, repairs and accessories and customer service with the fuel. Again, the use of well-known trade names helped guarantee the quality of both service and products.

In many ways, Canada became a nation structured on the cheaply-fuelled automobile in the decades following World War II. The shift to urban living, the unprecedented mobility of Canada's post-war population and the move toward shopping malls as centres of commerce are only three examples of changes the automobile has made in Canadian life.

The expansion of the automobile market guaranteed a certain volume of gasoline sales. However, to survive, marketers continually had to adapt to changes in the automobile, in people's lifestyles and in other economic conditions.

For example, the decades following the war saw technological improvements in the automobile. As a result, cars required less routine servicing, fewer oil changes and less frequent lubrication. The new automobile's complexity often ruled out repairs from the local service station. Full service stations began losing ground to operations better placed to intercept the suburban commuter. Competition within the industry forced marketers to consolidate. Gasoline price increases had made consumers price-conscious, contributing to the popularity of self-serve outlets and the decline of the smaller corner station.

The number of operating service stations declined rapidly in the 1970s, from nearly 36 000 to about 24 000. By the early 1990s, the figure had dropped to around 18 000. At the beginning of 1992, Imperial and Petro-Canada announced that they would close or sell another 2 300 service stations and distribution terminals. Thus, the trend continues.

In the 1970s, the decline in marketing outlets generally resulted from simple consolidation by the major oil companies. In the 1980s, however, other factors came to bear. Of particular importance were changes in corporate structure that came from a rash of takeovers and mergers. This had an immense impact on the industry, mostly visible to the public at the gas pump.

Until 1980, there were many marketing companies with regional interests, but four companies – Imperial, Gulf, Shell and Texaco – dominated the gasoline retailing business. This began to change when the federal Crown corporation, Petro-Canada, got into the marketing business. Petro-Canada did so by purchasing a string of companies with refining and marketing assets.[20] First it purchased Pacific Petroleums, then Petrofina, then BP's marketing and refining division. In 1985, it purchased most of Gulf's refining and marketing assets. Then Imperial bought up Texaco. Thus, the number of large integrated companies dropped to three – Imperial, Shell and Petro-Canada.

As these companies consolidated, they frequently found themselves with too many service stations in a single market area. At city intersections that once hosted Imperial, Gulf, Petro-Canada and Texaco service stations, for example, the new proprietors would sell or close the less profitable stations.

The sale of service stations once owned by large integrated companies contributed to the proliferation of independent marketers and so-called "discount" outlets. By 1992, independents, discounters and agents had boosted their market share enough to challenge the major marketers for supremacy. In an era of declining demand for refined products, these developments intensified competition in the gasoline marketing business. In the marketer's lingo, there was "too much product chasing too few cars."

THE BERTRAND ALLEGATIONS

Among many consumers there has long been a strongly-held belief that marketers somehow rig prices, gouging the public. This accusation made front-page headlines in early 1981. Robert Bertrand, then the federal Director of Combines and Investigation, charged the industry with price-fixing and other offenses.

Bertrand had begun his investigation in 1973, after receiving a complaint from the Consumers Association of Canada. In 1973 and 1974, the Combines Investigation Bureau searched the premises of 21 oil companies, seizing more than 200 000 documents. In 1975, 30 witnesses from oil companies appeared at a five week hearing. In 1976, Bertrand's office required 90 companies to file written reports. Petrofina challenged in court the bureau's right to search corporate offices, with the courts eventually ruling in Bertrand's favor.[21]

The issue simmered until Bertrand dropped his bombshell in 1981, releasing his charges in a public report. Briefly, he argued that between 1958 and 1973 the integrated companies had overpaid for imported oil. He claimed that the price of domestic oil refined in Ontario had been too high. And he charged that marketers had operated inefficiently, passing on their inefficiencies to consumers in unnecessarily high prices.

Bertrand assigned two price tags to each of his allegations. The second adjusted his numbers for inflation, suggesting a "present value" cost and further inflaming public outrage. The following table summarizes his contentions.[22]

PRACTICE	COSTS	
	Simple Total in 1980 Dollars	Present Value in 1980 Dollars
1. Inefficient gasoline distribution system	$5.2 billion	$34.3 billion
2. Import overcharge for crude oil	$3.2 billion	$28.2 billion
3. Import overcharge for distribution system	$0.6 billion	$4.1 billion
4. Excess cost of domestic crude in Ontario	$3.1 billion	$22.6 billion
Total:	$12.1 billion	$89.2 billion

Bertrand's charges set loose a national storm, with opposition politicians loudly proclaiming their indignation. Tory MP Allan Lawrence insisted that "the people have been ripped off by the oil industry; something must be done." NDP leader Ed Broadbent said Bertrand's report "documents the greatest rip-off in Canadian history."[23]

As he released his allegations, Bertrand did not call for criminal charges against the oil companies as he might have done. Instead he recommended further hearings by another federal agency, the Restrictive Trade Practices Commission. Already under pressure from the National Energy Program, an angry petroleum industry mounted a determined defense. In 1986 the commission released a three-volume report entitled "Competition in the Canadian Petroleum Industry."[24]

Based on an examination of 200 witnesses and thousands of documents and submissions, those volumes reported no evidence to support Bertrand's allegations. Wrote commission chairman O.G. Stoner, "the Director's case . . . was misconceived."

> There was no proof placed before the Commission that Canadian petroleum companies overcharged consumers by 12 billion dollars or that, indeed, any measurable excess costs were passed on in any significant degree between 1958 and 1973. Efforts by the Director [Bertrand] devoted to that piece of history could have been much more productive in examining current practices in the industry and would have shortened the inquiry.[25]

Although the commission's findings entirely exonerated the industry, its conclusions received little coverage in the press. Within the industry there was a sense of frustration that the front-page allegations of 1981 had unfairly blemished its reputation.

Partly because of this affair, in 1981 the federal government created the Petroleum Monitoring Agency to conduct annual and semi-annual reviews of the petroleum industry's financial performance.[26] Unprecedented in Canadian history, part of its mandate was to determine whether the petroleum industry was receiving "excess profits." Ironically, it did just the opposite. Although the graph line shows peaks and valleys, the agency essentially documented a steady slide in the industry's financial condition. Torn apart by intense competition, declining demand and two serious recessions, refining and marketing profits fell steadily during most of its twelve years of operation.

Chapter 11

Petrochemicals
(The Miracle Workers)

Novacor chemicals plant at Joffre, Alberta, 1984
(NOVA Corporation, 15801)

Petrochemicals are one of the great industrial triumphs of this century. It is difficult to imagine making a toothbrush or the body of a telephone, for example, out of anything but plastic. Nylon carpets are a part of home and business life. So are styrofoam cups. Athletes pulling on sports clothing, motorists using antifreeze, carpenters dabbing on glue, farmers spraying fertilizers on their crops rarely give a thought to the petrochemicals involved.

Through the alchemy of petrochemical processing, versatile substances like "alkyd" and "polycarbonates" become composite materials lighter and stronger than steel. They can increase the fertility of farmland a hundredfold. They provide insulation, pharmaceuticals, household cleaning products and literally thousands of other uses. To a degree that is difficult to comprehend, they make modern life possible.

The roots of the petrochemical industry go back to nineteenth century England and Germany, where chemists began developing dyes and pharmaceuticals from coal tar. Germany soon took the lead in organic chemistry, and by the turn of the century had developed advanced processes for using organic chemicals to develop fertilizers and nitrate explosives. However, the mantle of leadership in petrochemicals went to the United States in the twentieth century. According to one authority:

> Looking back now, it is difficult to select any single reason that would account for the dramatic growth of the U.S. petrochemical industry from a tiny base in the late 1920s. It was not the discovery of vast new sources of inexpensive hydrocarbons.... Nor was it a surge in the demand for products that could now be produced more economically from petroleum feedstocks, since the petrochemical industry gathered its momentum during the Depression years.... Favorable economics due to chemical engineering breakthroughs were probably a factor.... But in the final analysis, companies started to make petrochemicals because conditions were ripe for a change in the technology and feedstocks for the production of organic chemical intermediates. The management of several enlightened oil and chemical companies led this industrial transformation.[1]

The use of petroleum as a feedstock for chemical manufacture is strictly a twentieth century phenomenon. In the United States, the industry got its start during and after the First World War, as refiners found commercial uses for the gases generated by crude oil cracking. In some cases, refiners became petrochemical producers themselves; in others, they simply sold their off-gases to chemical companies. In addition, the petrochemical industry benefitted from new refining technologies, some of which found use in petrochemical production.[2]

From those beginnings the petrochemical industry grew rapidly. And its products became pervasive.

WHAT ARE PETROCHEMICALS?

The definition of petrochemicals is somewhat arbitrary. Sometimes it refers to substances as petrochemicals simply because they originated with oil and gas extraction – sulphur, for example, and helium. More commonly, however, the definition of petrochemicals refers to organic compounds created directly from oil and gas but not used for fuel or lubricants. Organic petrochemicals involve the versatile carbon atom, and are far more numerous than inorganic (non-carbon) petrochemicals. But the industry produces greater volumes of petrochemicals from such inorganic by-products of petroleum production as sulphur.[3,4]

At the start of the Second World War, the petrochemical industry was largely an American industry. The refinery cracking process, used to increase gasoline yields from crude oil, became widespread in the 1920s. One by-product was the family of simple olefins – ethylene, propylene and butene. Industrial chemists found ways to react olefins to form a series of alcohols and alcohol derivatives at prices competitive with those produced in the chemical industry from other starting points.

The next important petrochemical development was a series of olefin oxides which, unlike the alcohols, was new to the world. From olefin oxides came a variety of new products. These included chemicals used in the automobile industry, such as antifreeze and surface coating materials.

During the war the petrochemical industry became more of a global enterprise, although the United States still held a commanding lead. The petrochemical sector now used a broader range of hydrocarbons as petrochemical feedstocks: the paraffins, di-olefins, acetylene and the aromatics (hydrocarbons with ringed molecular structures).

WAR BABIES

The Canadian petrochemical industry was a war baby – not unlike the term "petrochemical" itself, which probably appeared for the first time in the June 1942 issue of *The Oil and Gas Journal*. The war was a direct stimulus to Canada's petrochemical industry; most of the explosives used by the Allied war effort came from the nitration of nitric acid and ammonia. To produce the needed ammonia, the Allied War Supplies Corporation built a plant at Calgary. This plant synthesized ammonia from natural gas piped in from the Royalite scrubbing plant in Turner Valley.

This plant, Alberta Nitrogen Products, produced 41 percent of Canada's total ammonia production. It was the only plant in the world at the time making ammonia from natural gas. The Allied war effort, however, needed food as much as it needed bombs. Later in the war the Calgary plant began producing ammonium nitrate fertilizer.

The second petrochemical war baby in Canada was the Polymer Corporation, a Crown-owned synthetic rubber plant stimulated into existence by Japan's 1941 attack on Pearl Harbor. Japan soon controlled over 90 per cent of the world's natural rubber production capacity. This was a serious concern, because in war even more than in peace, rubber is a strategic commodity. Much of the machinery of war cannot run without it.

Sarnia was chosen as the site of the synthetic rubber plant for four reasons. First, it was the site of the nation's largest refinery. Located only 35 kilometres from Petrolia, the Imperial Oil refinery is an inheritance from Ontario's early oil discoveries. Second, the St. Clair River offered a year-round supply of low-temperature water for cooling. Third, rail and water transportation facilities were available both to bring in raw materials and to take out manufactured chemicals. Finally, the site offered excellent rail and water transportation to Canada's industrial heartland.

Polymer was one of the great development projects of the war effort in Canada. Four large construction companies and ten leading engineering firms took part in the project, which had a peak labor force of nearly 6 000. Some 9 000 freight cars carried construction materials to the site.

Initially, three companies provided Polymer (later Polysar) with technical operating expertise. One was a subsidiary of Imperial Oil; the second, Dow Chemical of Canada; the third, a company funded by three natural rubber companies. The plant produced the first synthetic rubber ever made in Canada (and in the British Empire) in 1943.

Polymer continued operation after the war by aggressive marketing and product diversification, despite the resurgence of natural rubber on world markets. By 1947, Polymer was making solid rubber for products ranging from tire casings to footwear to foam rubber. A petrochemical success story, the company eventually owned plants in Canada, the United States and Europe.

Dow Chemical's participation in the Polymer plant played a role in petrochemical development at Sarnia. Dow began its own operations at Sarnia by converting Polymer's excess styrene monomers into the thermoplastic resin polystyrene. This was only the beginning of a rapid expansion of facilities by Dow at Sarnia. In time, the many plants near Sarnia – owned by Dow, Polysar, Shell, Imperial and others – earned that region the title of the "Chemical Valley of Canada."

After the war, Dow also began to take Polymer's excess ethylene, converting it to ethylene glycol, the key component of antifreeze. Because of its climate, Canada is a large consumer of antifreeze, but Dow was the first Canadian producer of ethylene glycol. And in 1947, the company began to crack ethane taken from plant off-gases to produce its own ethylene. It used this ethylene to expand its production of styrene monomer and glycols. Production facilities for chlorine and caustic soda and for ammonia followed. By 1954, Dow operated on 300 acres at Sarnia, employing 700.

In 1952, Cabot Carbon of Canada Limited added to the surge of products available from Sarnia by constructing a carbon black plant there. Used as a reinforcing agent for both synthetic and natural rubbers, Sarnia producers created carbon black by partially burning the heavy by-products ("bottoms") of the catalytic cracking process.

The development of Sarnia's petrochemical industry illustrates a common occurrence in petrochemical manufacturing. When one major petrochemical plant goes into operation, other petrochemical plants frequently locate nearby. The newcomers often use the petrochemicals or the petrochemical by-products produced by the original plant. They also take advantage of the skilled workforce and managerial expertise which developed at existing plants, and of such infrastructure as pipelines and delivery systems to petrochemical markets. In Canada this phenomenon has also occurred around Montreal and near the Alberta cities of Edmonton and Red Deer.

EDMONTON AND FORT SASKATCHEWAN: THE LEDUC EFFECT

If the Second World War was the primary stimulus behind the first petrochemical ammonia plant at Calgary and the first synthetic rubber plant at Sarnia, the next wave of development was the product of the oil plays in Alberta following Leduc.

The petrochemical industry has historically developed closer to its markets than to its feedstocks (assuming the two locations are far apart). There are two main reasons. First, the bigger markets have more sophisticated infrastructure, refineries and other chemical operations. Second, transporting raw materials in bulk to population centres helps reduce transportation costs for finished products.

Accordingly, most petrochemical plants constructed soon after the war clustered about the major central refineries at Sarnia and Montreal. The primary exceptions were the petrochemical fertilizer plants. These plants had the best of both worlds: by locating in the west, they were close to both the bulk of their consumers and to inexpensive supplies of natural gas feedstock.

There were, however, early exceptions of another kind: two important petrochemical plants constructed at the upstream end of the Canadian petroleum industry. The first of these was a cellulose acetate plant at Clover Bar, immediately east of Edmonton. Canadian Chemicals Company Limited, the Canadian subsidiary of Celanese Corporation of America, completed the plant in 1953, in its day the largest industrial undertaking ever in Alberta. It was similar to the parent company's cellulose acetate, yarn and fabric plant at Bishop, Texas. For feedstock, it used natural gas from the nearby Morinville field and alpha-cellulose from a Celanese-owned pulp mill in Prince Rupert, British Columbia. The plant was Canada's first producer of a wide range of petrochemicals.

The Edmonton area gas and oil finds of the late 1940s also brought Canadian Industries Limited (CIL) to Edmonton. The company located its polyethylene plant in Alberta in part because natural gas from Alberta's Leduc-Woodbend field was rich in ethane, the primary feedstock for ethylene production. So in 1953 CIL completed its plant in the Strathcona District of Edmonton on the North Saskatchewan River, close to the city's refineries. The plant extracted ethane from its feedstock gas stream, cracked the ethane to produce ethylene, and polymerized the ethylene to produce polyethylene. The polyethylene eventually found its way into such consumer and industrial products as packaging, pipe, housewares, kitchenware, toys, wire and cable sheathing, weather stripping and moulding.

MONTREAL: PROCESSING OFFSHORE OIL

For decades, Montreal had been a refining centre for imported oil. With the rapid growth of the petrochemical industry, the city's refineries became a catalyst for the growth of petrochemical plants and in 1953 several petrochemical facilities opened for business.

First, Shell Oil Company of Canada built isopropyl alcohol and acetone plants adjacent to its Montreal East refinery.

Next, BA-Shawinigan opened an acetone and phenol plant. This hybrid company was born of convergence of interests by two companies. British American (later Gulf Canada) wanted to use off-gases from its Montreal refinery for petrochemical manufacture. Shawinigan Chemicals wanted to convert its acetylene production from a limestone base to a more economical petroleum base. The new plant thus accommodated both companies' interests.

Then, Dominion Tar and Chemical Company (today known as Domtar) opened a plant in 1953. The facility produced ethylene oxide and ethylene glycol at Montreal, using as feedstock an ethane-rich fraction from nearby refineries. By the time the plant went on stream, increased production of ethylene glycol in the United States had resulted in heavy exports of the material to Canada at depressed prices. American exports to Canada arrived duty free, but the United States imposed a duty on Canadian exports at the American border. This created a difficult situation for Domtar, which soon sold its interest to Union Carbide Canada.

Oil-based petrochemical plants centred in Sarnia and Montreal proliferated and flourished during the 1950s and 1960s, as North America enjoyed a period of unprecedented prosperity. The two centres supplied large and rapidly growing markets for petrochemicals in Canada (southern Ontario and southwestern Quebec) and in the United States (the eastern seaboard and the Midwest).

Besides ready access to markets, these centres were near transportation systems which provided easy access to raw materials. Montreal area refineries and petrochemical plants relied upon overseas crude oil delivered by pipeline from Maine, while Sarnia used American feedstock. Sarnia began processing western Canadian oil supplies in 1952, when Interprovincial Pipe Line provided local refining and petrochemical plants with pipeline access to burgeoning Alberta supplies.

With federal subsidies, Interprovincial extended that pipeline to Montreal in 1976. This gave southwestern Quebec the opportunity to choose between low cost domestic crude oil and more expensive offshore oil. After the deregulation of oil prices, however, Montreal began finding Canadian oil more expensive than oil from offshore. By 1992 there was talk about putting the pipeline in mothballs, and Montreal's petrochemical industry relied heavily on offshore oil.

PETROCHEMICALS IN THE WEST

Unlike the oil-based petrochemical industry in central Canada, Alberta's natural gas-based industry saw relatively little growth after the early 1950s. The synthetic fertilizer business grew because of nearby agricultural demand for its product. Otherwise, Alberta's petrochemical industry remained roughly the same size for nearly two decades.

In the early 1970s, however, Alberta's petrochemical industry began to grow rapidly. One cause was the OPEC oil price shock of 1973, an international crisis which created a climate of real concern about the reliability of oil as a petrochemical feedstock. Oil had suddenly become costly and supplies were at the mercy of the unstable politics of the Middle East.

As a result, petrochemical producers began to look on natural gas much more favorably. This was partly due to regulated prices – it was relatively less expensive than oil – and the fact that production in Canada was high. Simply switching to natural gas also carried risks, however. There was a strong belief at the time that Canada and the United States would face natural gas shortages by the end of the decade.

While discoveries had increased reserves almost every year since 1947, the demand for natural gas surged in the early 1970s. If the growth in gas demand continued unabated, demand would quickly outstrip supplies. In the crisis-ridden climate of the 1970s, conventional wisdom held that natural gas shortages were imminent.

Manufacturers believed one way around this dilemma was to locate petrochemical facilities in Alberta. This made economic sense for two reasons. One was that locating within the province would provide petrochemical producers with secure access to natural gas feedstocks. Natural gas feedstocks include natural gas proper (chemically known as methane, or CH_4) and butane (C_4H_{10}), propane (C_3H_{10}) and, of particular importance, ethane (C_2H_6). Alberta's natural gas processors strip ethane from natural gas. The ready availability of ethane in particular was an important factor in the development of Alberta's petrochemical industry.

The other incentive to locate in Alberta was that natural gas prices were low. Before gas deregulation began in 1985, prices in Alberta were lower than federally regulated prices for gas sold outside the province. After deregulation, when prices began to result from negotiation between buyer and seller, Ontario and Quebec had to pay premiums for western natural gas because of the cost of transportation. And even secondary plants in Alberta that do not use ethane as a raw material as a feedstock found advantages in the low cost of gas, since they use that fuel for such manufacturing processes as steam cracking.

As it began to make sense to manufacture petrochemicals in Alberta, a blizzard of proposals hit the boardrooms. Some went before provincial regulatory bodies, others went no farther than the business pages of the press. But by 1980 dozens of companies had proposed new plants.

Although the provincial government provided few direct incentives for development, it wanted petrochemical jobs and investment in Alberta. Part of Alberta's diversification strategy was to encourage petrochemical development so the provincial economy could enjoy the value added to natural gas feedstocks by petrochemical manufacturing. Since Nova, Dome Petroleum and Dow Chemical all entertained plans to convert Alberta ethane into ethylene in Alberta, it looked as though at least one of these would be the bell cow for a herd of ethylene-based secondary petrochemical plants.

Nova (then known as Alberta Gas Trunk Line) was eager to diversify from its original gas-gathering function, and in 1974 provincial legislation enabled the company to do so. With encouragement from the province, the company proposed an ethylene plant at Joffre, a village near Red Deer, Alberta.

Dome Petroleum and Dow Chemicals proposed a separate scheme which had an ironic twist. At Sarnia, several local operators (including Dow) had joined in a project to make ethylene based mostly on propane and butane feedstocks. The project went by the acronym SOAP (Sarnia Olefins and Aromatics Project). SOAP considered both Alberta ventures as competitors. Dow thus found itself in an odd position, having also agreed to join Dome's Alberta scheme. The SOAP group of companies formed Petrosar Ltd. that year and Petrosar completed its Sarnia ethylene plant in 1978.

Dow eventually dropped out of SOAP and Dome dropped out of the Sarnia ethylene project. Novacor Chemicals, a Nova subsidiary, took over both ethylene projects. Two ethylene plants thus ended up at Joffre, both served by a pipeline that gathered ethane from both straddle and field gas plants. The second Alberta Gas Ethylene plant at Joffre came on stream in 1984.

By the time the first wave of secondary petrochemical operations hit Alberta, it had lost much of its drive. As world economic conditions changed, including a severe recession in the early 1980s, many project sponsors cancelled their plans. But some producers kept faith with the province.

Dow led the way, expanding into ethylene oxide, ethylene dichloride and vinyl chloride at Fort Saskatchewan. Novacor built a linear low density polyethylene plant at Joffre, using a process developed by Union Carbide. Other petrochemical plants based on Joffre ethylene also went into production. Also, Shell's new synthetic oil refinery hosted a styrene monomer plant.

Besides the ethylene-based petrochemical plants at Joffre, western Canada has methanol and ammonia facilities which use natural gas feedstock. Two of the ammonia plants are in Medicine Hat, Alberta; others are in Edmonton and Kitimat, British Columbia.

Two later developments hindered Alberta's development as a major petrochemical manufacturing centre on the scale foreseen in the 1970s. First, new discoveries combined with conservation eliminated the threat of shortfalls of natural gas. In fact, by the early 1980s the industry could not sell all the gas it could produce – a problem which refused to go away for more than a decade. Second, surpluses of oil began to develop on world markets. Taken together, these conditions eliminated concern about feedstock shortages and thus undermined some of the rationale for locating in Alberta.

However, Alberta's petrochemical industry had developed momentum, and by the 1990s offered several key advantages that kept it growing and competitive worldwide.

One advantage was that natural gas feedstock remained relatively cheaper than the crude oil feedstock used by many competitors. Also, Edmonton was western Canada's principal refining centre, and off-gases and other refinery by-products were used in petrochemical manufacturing. In addition, Alberta had readily accessible and easily transported supplies of both ethane and butane, as well as new, large-scale and efficient plants.

But at least as important was the phenomenon of synergy. As happened in Sarnia and Montreal, petrochemicals and by-products from one plant encouraged the development of other plants nearby. In addition, petrochemical infrastructure developed for one plant could often be extended to service new petrochemical ventures.

The creation of the petrochemical service and supply sector and of a skilled workforce were also factored into Alberta's success. So was the concentration of managerial expertise, which gained particular momentum in the 1980s.

THE NOVA FACTOR

A hint of building momentum in the petrochemical industry came in 1984, when Shell Canada began moving its head office from Toronto to Calgary in conjunction with the opening of a styrene plant and refinery in the small prairie town of Scotford, near Edmonton. While these downstream factors were important, Shell's main motivator was the growing importance of its oil and gas business.

Nonetheless, Shell's standing as one of Canada's largest petrochemical operators made a difference. Initially, the company's head office for chemicals remained in Toronto, close to customers in central Canada and the United States. In 1990, however, Shell formally moved the division to Calgary, noting the economies of consolidating its chemicals headquarters with corporate head office and the need to be closer to the company's products management team. Also, rapid globalization of the world economy meant that Shell was developing more major overseas customers and had less need to locate in central Canada.

Shell's move was a symbol of Alberta's strengthening petrochemical muscle. It meant the most senior executives of this highly successful company would have a clear view of provincial opportunities. It also meant a gathering of technical and managerial expertise.

By the late 1980s, there were as many large petrochemical plants in Alberta as there were in Ontario, twice as many as in Quebec.[5] Although Ontario had more sophisticated chemical expertise, the prairie upstart was closing in.

Then NOVA Corporation took a bead on Polysar, the grandfather of central Canada's petrochemical industry. In October 1987, the Alberta corporation bought 6 million common shares of Polysar stock on the open market before announcing that it wanted to acquire Polysar completely.

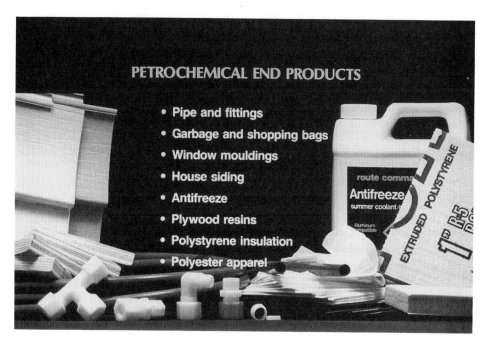

Examples of products made from petroleum
(NOVA Corporation, 16747)

Polysar fought the takeover, and the two companies engaged in high-profile corporate war. Nova won, but the battles cost it dearly. Nearly a year after its initial stock purchase, the two companies merged. Nova paid the equivalent of $1.9 billion, far higher than its original bid.

With this purchase, Nova became the largest petrochemical manufacturer in Canada, the sixth largest in North America and one of the 20 largest in the world. The company's standing among the world's top-ranked petrochemical companies lasted only two years, however. In 1990 it sold the rubber division – ironically, the progeny of Polysar's original wartime mandate – to a German company. At the exchange rates of the day, the price was $1.25 billion.

Notwithstanding its sale of assets, Nova remained Canada's largest petrochemical producer. And its presence in Alberta, like Shell's, is likely to help drive the momentum and synergy which helped build large petrochemical operations around the province.

EXPANDING MARKETS

Unlike the refining and marketing businesses, petrochemical production and sales have grown steadily since the sector's early years in Canada. Although the Quebec-based industry is in decline, all three Canadian petrochemical centres – Alberta, Ontario and Quebec – continue to produce large volumes of petrochemicals. The aggregate production from the three centres has made Canada's petrochemical industry the tenth largest in the world, based on the value of production.

There were several reasons for the industry's continued growth: industrial chemists created new petrochemicals and industry found new applications for them; the demand for chemicals tends to grow with a growing economy; international demand expanded rapidly, creating export opportunities.

In constant dollars – that is, dollars which fully reflect inflation – Canadian petrochemical sales increased by more than half in the decade beginning in 1982. To a large extent this reflects the impact of exports, which rose from 29 percent of sales to 42 percent during that period.[6] The difference in export demand came from countries other than the United States, notably the Pacific Rim. This is ironic, since the Free Trade Agreement between Canada and the United States began eliminating tariffs in 1989 (the last disappeared at the beginning of 1993). Yet during the first three years of free trade, Canadian exports to the United States remained constant at about 57 percent of total exports – unchanged from 1982 levels.

As earlier chapters have shown, the Free Trade Agreement was only one of many government policies which affected the petroleum industry and its offspring. The next chapter takes a deeper look at the interplay between policy and industry.

Novacor chemicals polystyrene facility in Montreal produces base material for many plastic products, 1992 (NOVA Corporation, 335822)

Chapter 12

A Matter of Policy

Hibernia gravity-based system under construction in Newfoundland, 1993
(Petro-Canada)

Politics proved to be as tough as the geological and geographical challenges faced by Canada's oil industry. An incident in Ottawa on the morning of December 7, 1979, illustrated the power of political forces with startling clarity. It was one of the brief dramas of the daily question period in the House of Commons that occasionally serve as its version of pictures worth a thousand words.[1]

OIL POLITICS TOPPLED A GOVERNMENT

Joe Clark and Ed Lumley took centre stage. Within a year, Lumley was trade minister in a Liberal government that Western Canadians reviled for introducing the National Energy Program (NEP) and its gas price regulations, export controls, special taxes and investment rules. But in this exchange, Lumley was a Liberal opposition MP from eastern Ontario who only had power to ask questions. Clark was the one in a position to make a difference.

On that Friday morning, Clark stood out as an Alberta-born prime minister who led an avowedly pro-business Conservative party and represented a constituency studded with oil and gas installations in the foothills of the Rocky Mountains near the industry's home-base, Calgary. When Lumley asked if the Tories would respond to reports that Canadians arms of international oil companies raised prices to enrich their foreign parents.

Even this prime minister registered the power of the consumer side in national politics. Clark said:

> *We are now actively discussing the mechanism that will be used to ensure that virtually 100 percent of the revenues that would go to the companies as a result of the increases in energy prices will, in fact, be regained by the Government of Canada for the specific application by the Government of Canada for national energy purposes.*

With four hours the Toronto Stock Exchange's oil and gas index fell 107.29 points even though Clark's aides scurried to add explanations and qualifiers meant to take the sting out of the remark. Securities analysts said the market was jittery anyway. Clark only confirmed speculators' suspicions that it was time for some "profit-taking" on stocks they had bid up in response to oil price increases, which were in turn owed to similarly speculative behavior on commodity markets animated by the continuing revolution in Iran.[2]

In case there were any doubts that Clark meant what he said, he stuck to his guns when the issue was raised again in Parliament the following Monday. This surprised no one who paid attention to energy affairs. Clark was in his fourth month of embarrassing, high-profile failures to reach a federal-provincial agreement on energy prices and revenue-sharing with two fellow Alberta Tories: Peter Lougheed, the premier, and his energy minister Merv Leitch.

GOVERNMENTS PLANNED ON ESCALATING OIL PRICES

Clark's reply to Lumley only stated the federal approach in a nutshell. The Tories' proposals included government management of the rate of price increases, an "energy self-sufficiency tax" to capture for Ottawa $2 billion a year in the resulting "incremental" or new revenues, a "Canadian energy bank" to direct this money into supply developments favored by the government, and rules for using this money that would favor small companies – meaning Canadian ones – over the international corporations. The whole scheme relied on forecasts that world oil prices would keep on climbing, to more than double the $42 US a barrel by 1985.

While differing in details of numbers and program names, this was in broad outline the later National Energy Program. Lougheed and Leitch certainly saw matters that way. They fought Clark so hard that his minority government was only able to announce a six-month extension to its original target date for an agreement when its finance minister, John Crosbie, presented the budget that was defeated in the Commons within a month of the exchange with Lumley.

The national Tories' defeats in the Commons, then in the 1980 election, turned at least partly on energy aspects of that budget, including Clark's troubles persuading Lougheed to co-operate with his "national energy purposes." The NDP's leader, Ed Broadbent, made energy his main issue for the campaign. The Liberal's Pierre Trudeau gave energy room on his priority list too, along with their mainstays of national unity and constitutional reform.

Energy issues come naturally to Canadian politics. With a thin population stretched over the world's second-largest country after Russia, a climate often as chilly, an economy heavily reliant on mechanized resource extraction and the personal income to demand creature comforts, Canadians stand out as the highest per-capita users of energy on earth.

Tension over the use, pricing and proceeds of energy production also comes naturally, thanks to the federal constitution that makes this vast country possible. Lougheed acknowledged that this tension over Ottawa's control of natural resources is no mere matter of personalities or even differences of political emphasis on state management versus free enterprise.

PRODUCERS AGAINST CONSUMERS – PROTECTIONISM AND SUBSIDIES

Even during their worst disagreements, Lougheed held no personal or ideological grudges against Clark. Instead, Lougheed cited his fellow Albertan's positions and headaches in Ottawa as illustrations of his reasons for turning down invitations to run for the leadership of the national Conservative party in 1976, leaving the race open to Clark. "I could see this coming" Lougheed said: "The confrontations would get more extensive. The West would have to be more determined."

Habits of demanding government protection have been ingrained for most of Canadian history among oil and natural gas consumers, who in the beginning, as now, elected the majority of members of Parliament.

Discoveries of natural gas in Ontario in 1889-90 led swiftly to the first Canadian exports to the United States, via a pipeline to northern New York. By 1907, it became obvious that initial expectations of inexhaustible supplies were wrong. Ontario consumers who had switched to the new fuel mounted an outcry against the exports. The provincial government responded by passing an Exportation of Power and Fluids Act, which restricted exports to only volumes in surplus of domestic requirements.

The Canadian tradition of government help for the supply side of the petroleum industry is just as old. This involvement has included precursors to the NEP's grant scheme, the Petroleum Incentive Program, and Petro-Canada. For example, between 1904 and 1925, the federal government paid a bounty of up to 52 cents per barrel to the operators of Canada's first oil fields around Petrolia and Sarnia in southwestern Ontario. From 1943 to 1945 a federal Crown corporation, titled Wartime Oils Ltd., bankrolled drilling in Alberta with loans made on lenient terms.

The British North American Act, which made Canada a country in 1867, guaranteed there would be continuous, complex and often volatile interplay over energy among consumers, producers, federal and provincial authorities.

The constitution launched this intricate dance of economic and political forces by creating a division of powers that reflected a desire for a strong central government along with self-supporting provinces.

The BNA Act assigned provincial governments 16 responsibilities, such as education and property and civil rights. To help pay for services needed to fulfil these responsibilities, the provinces were given ownership of natural resources within their boundaries as sources of economic activity and "rent" payments such as production royalties.

The powers of the federal government were listed under 29 headings and included a "residual" clause that made Ottawa responsible for all matters not specifically assigned to the provinces. Of particular importance to the development of energy policy, the federal government retains authority over trade and commerce across provincial and national boundaries, plus the right to levy both direct and indirect taxes.

Canada barely made in into the teen years before this constitutional arrangement showed its potential to be a recipe for conflict. The division of powers rapidly turned out to be a fertile ground for seeds of regional conflicts that were sown by founding prime minister John A. Macdonald's National Policy.

THE NATIONAL POLICY: HINTERLAND SUPPORTS METROPOLIS

This policy aimed to build a nation by supporting the industrial heartland with protective tariffs while settling the West via the Canadian Pacific Railway. The policy worked so well that the west rapidly filled up, developed a political consciousness, and sprouted lively populist movements dedicated to resisting the region's assigned role as the provisioner of the centre.

A healthy share of the agitation centred on reversing Ottawa's apparent aim to keep the West a hinterland, by refusing to extend resource ownership to Manitoba, Saskatchewan and Alberta when they became provinces in 1871 and 1905. The Western provinces only took their resource from Ottawa after tremendous political campaigns and a ruling in 1930 by the Privy Council of Britain's House of Lords. The marathon effort left an enduring legacy of distrust of central Canada and a determination among Westerners to guard their resource rights fiercely against any federal action that might diminish them.

RESOURCE CONSERVATION ISSUES

Tension between governments and the petroleum industry likewise dates back to the dawn of oil and gas development. Resource conservation was one of the first issues. In Ontario's nineteenth century oil fields and during the first two decades at Turner Valley, the industry regarded natural gas as mainly a nuisance by-product and simply "flared" or burned it.

In the very early years, Ontario's Tillbury producers even believed that to extract oil from a well where gas was also present it was essential to get rid of the gas first. This version of reservoir mechanics was dead wrong, and let to terrible waste of gas and oil. The Ontario government forced a change in practices to conserve gas around the turn of the century.

Early experiences with natural gas in Alberta involved little waste. The first gas fields were developed exclusively to supply utilities: Medicine Hat to serve its namesake city in 1900; Bow Island for Lethbridge, Calgary and other southern Alberta points in 1912; and Viking, which began to supply Edmonton in 1920. Only at Turner Valley, an oil field, was gas wasted. Driven by high underground pressures, the gas at first came up from the reservoir with such explosive force that it seemed to be a limitless resource.

The first significant act of resource conservation in Alberta came in 1930. Canadian Western Natural Gas reinjected Turner Valley gas into the Bow Island reservoir, where depletion triggered popular demand for conservation to protect supplies for southern Alberta communities.

The project did not end the waste of Turner Valley gas, however. In the 1920s and well into the 1930s, the object of producing this "wet" gas was its content of naphtha or natural gasoline. Only one producer, Royalite, had a market for its residue of natural gas through a pipeline to Calgary. Others disposed of it as an unwanted by-product. The waste was encouraged by a widespread belief that the reservoirs were under such high pressure that the gas flows could not be safely choked off. Production was wide open; sky-high flares that could be seen from Calgary and many other southern Alberta communities became part of Alberta folklore and even a tourist attraction.

FINDING MARKETS PIPELINE STRATEGIES

A variety of ideas emerged for using Turner Valley gas, including the construction of pipelines to Regina, Winnipeg and Toronto. Others proposed to extract methane and ethane. The federal government, which retained jurisdiction over Turner Valley until the transfer of resource ownership to Alberta in 1930, had considered plans to reinject the gas into the reservoir, export it or use it in a carbon-black plant. The onset of the Great Depression killed these plans.

The federal government also attempted to control the waste. A 1922 regulation restricted drilling, while staff of the Department of the Interior urged operators to use proper well practices, collect samples and keep production records. Stronger initiatives were in preparation when resource jurisdiction passed to the provincial government in Edmonton.

The extent of the waste was documented by the Alberta Public Utilities Board in 1931, during hearings on a request by the City of Calgary for reduced gas rates. By then, it was estimated that about seven billion cubic metres of gas had been wasted for a loss of one-third to one-half of the field's reserves. The figures shocked provincial authorities and the Alberta government rapidly took steps to introduce conservation.

Well-pressure tests demonstrated that the gas flows could be safely shut off. The government, then the UFA or United Farmers of Alberta, enacted legislation creating the first provincial oil field watchdog agency in 1932. Titled the Turner Valley Gas Conservation Board, its mandate was to seize control of the chaotic field by rationing production fairly among its numerous producers.

The board had to face up to a tough problem. Only the larger companies could afford to defer income over the short term, in the name of conservation, in order to reap greater gains from keeping the gas until it could be sold in the longer run. Smaller producers relied heavily on regular monthly incomes. The board was told that production cuts to achieve conservation would in many cases threaten companies with bankruptcy. One such company eventually appealed to the courts against an order to cut production by 95 percent.

The Alberta government won the case in the Supreme Court of Canada. But the victory was hollow, because the court ruled that production cuts could not be retroactively imposed on mineral leases issued before the 1930 transfer of resource jurisdiction. With equal application of the conservation policy thus made impossible at Turner Valley, the board and the production controls were soon abandoned. Requests for voluntary improvement to production practices only scored partial success.

The true extent of the resource wastage at Turner Valley only became known after 1936, when deeper drilling revealed an oil reservoir below what turned out to be only a "gas cap" on a much bigger formation. The depletion of the gas cap deprived the deeper reservoir of much of the natural "drive" or pressure that could have driven oil to the surface. So the field's productive capacity was badly damaged by the time the oil was discovered.

Quick action was needed to prevent further waste of the gas cap. At the same time, the new find was big enough to cause surpluses. Producers had trouble finding markets for all the oil suddenly available. Both headaches could be cured by production cutbacks, and these were with co-operation from the federal government.

In 1938, Parliament passed a Resource Transfer Act to affirm that the Alberta government also received control over federal production licenses that predated the jurisdictional change in 1930. At the same time, the provincial government moved quickly to use the new authority by establishing Alberta's second oil field agency, the Petroleum and Natural Gas Conservation Board. This time, the watchdog survived and acted swiftly to control well spacing, rates of production and gas flaring. During the Second World War, the board succeeded in enforcing a scheme to gather all Turner Valley residue gas into a treatment plant. Unsold gas had to be reinjected into the gas cap.

Later renamed the Energy Resources Conservation Board (ERCB), this Alberta agency has become a model regulator with numerous imitators abroad, by consistently limiting the flaring of gas and sulphur by-products ever since its birth.

In creating the ERCB, the Alberta government also made one major departure from tradition established in the United States where members of the granddaddy of oil field regulatory agencies, the Texas Railroad Commission, are elected. ERCB members are appointed for lengthy terms. To further free it from political pressure, the ERCB was set up as a "quasi-judicial" agency akin to a law court, clearly separate as "a body politic and corporate" in its own right rather than a government department. The head office was located in Calgary rather than in the provincial capital of Edmonton. As the ERCB gradually expanded its authority to coal mining and electricity as well, the agency established a practise of drawing its budget from equal contributions by the government and the industries being regulated.

Provincial policy reached beyond just conserving resources to try to ensure more would be developed, as soon as the discovery of oil in Leduc in 1947 proved new supplies could be found. Within a year, Alberta's Social Credit government of the era set a ceiling of 16.67 percent on royalties for production from Crown-owned mineral rights, or more than 80 percent of the provincial resource endowment. To lure industry with a pledge of stability, the ceiling was set for the life of each petroleum and natural gas lease. The pledge held for the lifetime of the Social Credit government until 1971. Saskatchewan followed through in 1953, with guarantees that its royalties would hold steady for 10 years at 16 percent for oil and eight percent for gas.

While oil held the popular spotlight, the gas side of the industry heated up. Confronted with entrepreneurial proposals for pipelines to take gas beyond Alberta's borders, the provincial government called an inquiry into reserves and named it the Dinning Commission.

The inquiry's results echoed the industry's earliest experience in Ontario in 1907. After hearing testimony that gas was so important that a 50-year supply should be reserved for Alberta needs, the commission recommended a priority list that both the provincial and federal governments eventually adopted. The policy principle held that Albertans rated first call on provincial gas supplies, followed by other Canadians call over foreign customers.

The Alberta government gave the ERCB responsibility to hear all requests for permission to remove gas from the province. A legislative response to the Dinning Commission, the Gas Resources Preservation Act, gave the board a mandate to hold back a 25-year supply for all foreseeable Alberta needs.

RESOURCE INCOME SHARING

The federal government likewise responded promptly to the boom triggered by the Leduc discovery. In 1949, Parliament passed an Oil or Gas pipelines act to police transportation of oil and gas across provincial and international boundaries. The rules required builders of long-distance pipelines to be incorporated by acts of Parliament, and to submit their projects to the Board of Transportation Commissioners.

Aggressive drilling triggered by Leduc found enough gas to persuade the governments to allow exports. The trade opened in 1952, with construction of Canadian-Montana Pipeline to supply a strategic metals milling plant that made supplies for the American-led "police action" to defend South Korea against invasion by North Korea and China.

In these early years of the modern industry, the federal government matched its provincial counterparts as a promoter of growth. Within a year of election, John Diefenbaker laid to rest any notion that his national Conservative government opposed development. This idea may have lingered after the party's bitter fight, as the opposition side in the famous pipeline debate, against the former Liberal administration's methods of supporting construction of the TransCanada gas system.

THE NATIONAL ENERGY BOARD AND THE NATIONAL OIL POLICY

The Diefenbaker cabinet delivered a highly pro-development policy. This was an attempt to cure supply gluts that resulted from a temporary shortage caused when the 1956-57 Suez Crisis accelerated drilling that had already been booming in the West ever since Leduc. An inquiry, called the Borden Commission after Chairman Robert Borden, declared in 1958 that further gas exports should be allowed because Western Canada's reserves far exceeded the nation's foreseeable needs. Within a year, Alberta authorized increases in gas removals from the province. Part of the new exports launched development of the province's biggest export market: sales to San Francisco's Pacific Gas & Electric Co., with supplies bought by procurement subsidiary Alberta & Southern Gas Co. and delivered through pipeline subsidiary Pacific Gas Transmission Co. Shipments commenced in 1961.

In Ottawa, Parliament carried out another Borden recommendation – which had also been foreshadowed by a similar suggestion from a parallel Gordon Royal Commission on Canada's Economic Prospects – by creating the National Energy Board (NEB) in 1959. In the co-operative style then in favor in Ottawa, the federal cabinet reached out to Alberta to recruit Ian McKinnon as the NEB's first chairman. He had served 10 years as chairman of the provincial ERCB. The NEB rapidly adopted the Alberta principle of allowing exports of gas found to be surplus to a 25-year supply for Canadians, although a different formula emerged for making the calculation.

A second Borden report in 1959 turned to promoting oil development with a declaration that conditions were ripe for Canadian production to reach 110 000 cubic metres a day for sale throughout North America. The commission requested federal help for the industry to reach the target, in the form of import restrictions to reserve Ontario for oil from Western Canada. The recommendation was enacted in 1961 by the Diefenbaker government's National Oil Policy (NOP).

To meet ambitious production targets of 99 000 cubic metres daily by 1961, then 127 000 by 1963, the NOP sanctioned exports to the United States and created an institution called the Ottawa Valley Line. East of the line, Montreal and Atlantic Canada continued to import crude oil but refineries everywhere else had to use Western Canadian supplies. The policy held despite protests from Ontario, where it was realized that imports were cheaper, and from Home Oil, which had plans to build a pipeline to supply Montreal from Alberta.

NATIONALISM: CONSERVATIVE, LIBERAL AND INTERNATIONAL

While the NOP nurtured development for more than a decade, these years also saw the beginnings of the energy-policy instability that so dramatically marked even Joe Clark's Tory administration. At the same time as both Ottawa and Alberta began working on opening up geological and technical frontiers in the oil sands, the Arctic and the East Coast offshore, the first stirrings of political anxiety over foreign ownership emerged.

The Gordon Commission, named for renowned economic nationalist Walter Gordon, went beyond the idea of an NEB to make an array of proposals for raising the petroleum industry's Canadian content, ranging from mandatory ownership levels for companies to compulsory reliance on Canadian equipment suppliers.

This nationalist flavor repeatedly surfaced in actions ranging from the creation of the Canada Development Corporation to "buy back" assets from foreign owners to the establishment of Petro-Canada. Although this spirit reached its height under Liberal governments, the Diefenbaker administration responded too with a declaration of nationalistic intentions. Its 1961 edition of regulations for exploration and development on federally-controlled northern and offshore territories placed restrictions on granting production licences to foreign-owned companies.

The nationalist trend outlived the Diefenbaker government, to the Liberal administration under Lester Pearson, which in 1963 proposed new tax laws intended to raise Canadian ownership of oil and gas companies and block foreign takeovers. Although the Pearson cabinet listened to critics and withdrew the proposals, it flew the flag over the oil industry in 1967 by acquiring a 45 percent share in Panarctic Oils, a consortium drilling the far North. Besides keeping costly frontier exploration going, federal involvement was meant to affirm Canada's territorial claims over the Arctic islands.

THE CRISIS OF THE EARLY 1970s

While comparative peace over energy policy prevailed in Canada under Ottawa's NOP and the provinces' reliable royalty regimes, an end to stability was brewing as more aggressive versions of the same economic and nationalist aspirations came together in more volatile places. Nine years after 13 countries with more recent colonial and revolutionary pasts formed the Organization of Petroleum Exporting Countries (OPEC), the cartel started growing teeth. OPEC's avowed goal was to assert control over their oil and revenues. In 1969, revolutionary Colonel Moammar Ghadafy set a new tone by following through on his overthrow of King Idris with actions that forced international oil companies to pay OPEC member Libya higher prices and taxes.

Inspired by Ghadafy's successful show of strength, OPEC meetings in Caracas in 1970 and Teheran in 1971 generated agreements on solidarity and joint action to raise government participation in the oil industry, along with prices and taxes. A new view of the globe's natural resource endowments as limited stoked anxiety caused by OPEC's development. An international alliance of conservationists, The Club of Rome, won global attention with a prediction that expanding world population and industrialization would exhaust the planet's fossil-fuel reserves early in the 21st century.

Canada echoed OPEC's examples and the club's fears. In 1970, Quebec created a provincially-owned petroleum company called SOQUIP. A year later, the Gordon Commission's nationalist flavor found practical expression with the creation of the Canada Development Corp., to "buy back" Canadian industries and resource with deals that included a takeover of the Western operations of France's Aquitaine and their conversion into Canterra Energy.

Also in 1971, the federal government blocked a proposed purchase of Canadian-controlled Home Oil by American-based Ashland Oil. The new wave of direct action spread to Alberta when Premier Lougheed and his Conservatives won power in 1971 by ending 36 years of Social Credit rule. Lougheed's elaborate election platform, titled New Directions, sounded themes common among OPEC countries by pledging to create provincial resources and oil growth companies, collect a greater share of energy revenues, and foster economic diversification to prepare for the day when petroleum reserves ran out.

The idea of limited resources emerged from the realm of theory into hard facts of policy when the NEB rejected natural-gas export applications in 1970 and 1971, on grounds that there was no surplus and Canada needed the supplies. The strength of the new conservationist sentiment was underlined when the NEB stuck to its guns despite a 1971 declaration by the federal

Department of Energy that it thought Canada had a 392-year supply of gas and enough oil for 923 years.

In Alberta, government reserves estimators worked on predicting times in the foreseeable future when conventional oil and gas supplies would run out and the province would have to rely on oil sands plants and new industries. In their first session of the provincial legislature as a government, the Alberta Conservatives acted on their election promises in the spring of 1972. Royalty increases were proposed to raise revenue for industrial diversification. Over vehement objections from oil and gas leaders during rare public hearings held by the Legislative Assembly sitting as a commit, Alberta replaced its 16.67 percent royalty with a 25 percent rate for new leases. Saskatchewan set a new level of 24 percent, while its more radical brother New Democratic Party administration in British Columbia established a maximum of 40 percent.

Ottawa soon responded to world oil-price increases by a steadily tougher OPEC, rising American demands for Canadian supplies and forecasts that Alberta and international companies would siphon off the nation's wealth. Central Canadian critics soon invented a new nickname for Albertans – "blue-eyed Arabs." As prime minister of a minority Liberal government, Pierre Trudeau fired the first shot in what became known as the "70s energy wars" on September 4, 1973. He requested a voluntary freeze on oil prices. Nine days later, his administration imposed an export tax of $2.52 per cubic metre on oil. This marked the birth of an intricate system for enforcing and financing a "made-in-Canada" oil price that lasted 12 years. The tax equalled the difference between Canadian and international oil prices. The revenues were used to subsidize eastern refiners' purchases of imports.

Alberta responded October 4, 1973, by cancelling its 1972 royalty package in favor of a flexible system tied to changes in international oil prices. Two days later, the Yom Kippur War broke out between Israel and the Arab states. OPEC took advantage of the conflict to accelerate price increases by nearly doubling its posting for benchmark Saudi Arabian light oil to $32.30 US per cubic metre. The acceleration of international price increases, coupled with fears of shortages that briefly came true in the United States, rapidly aggravated tensions among provincial, federal and industry leaders. The rest of the 1970s were marked by rapid-fire, escalating actions and countermoves, as federal-provincial agreements alternated with renewed conflicts ignited by international developments.

On the provincial side, the B.C. Petroleum Corp. was made the province's sole buyer of natural gas. Saskatchewan set up a Crown energy company, Saskoil. Saskatchewan also passed legislation to appropriate all proceeds from oil price increases. This act, appealed by an alarmed industry, was struck down by the Supreme Court of Canada as a tax grab in all but name. The landmark ruling, called the CIGOL case after the company that started it, rendered the provinces more cautious but did not stop their flood of actions.

The mid-seventies also saw the birth of Alberta Energy Co., with the provincial government initially owning 50 per cent and endowing its creation with rich natural gas reserves and a role in oil sands development. The Alberta Petroleum Marketing Commission (APMC) was established to affirm provincial resource ownership. On paper, APMC took possession of all Alberta oil between its production and sale. In practise, APMC developed a strong role in the distribution of natural-gas revenues under a system called the "export flowback." Throughout the energy wars, Ottawa refrained from levying an export tax on gas even though its price on domestic markets was controlled. The flowback parcelled out to all producers, whether exporters or not, shares of revenues owed to higher prices in the United States. An old rule of politics prevailed, however: while one hand gave, another took away. Alberta set a 65 percent royalty rate on new revenues owed to gas-price increases, and Lougheed advised Ottawa he intended to do the same with oil.

On the federal side, the NOP died in 1973 when the Trudeau government announced the first in a series of new national policies. The new policies were founded on revenue-sharing, made-in-Canada pricing, increasing Canadian ownership of the industry and a quest for complete self-sufficiency in oil supplies through development of the oil sands, the Arctic and offshore potential. Petro-Canada was born, and so was the Foreign Investment Review Agency (FIRA). A 1974 federal budget countered rising provincial royalties by declaring them to be no longer deductible for income tax purposes, ending a privilege held by the industry since the 1940s.

Lougheed raised tempers to new highs with his reaction to the tax change, which was kept by a new federal budget after the first version was defeated; the Liberals won a majority government in the ensuring election. Lougheed called the affair "the biggest rip-off of any province that's ever occurred in Confederation's history." By 1975, tempers cooled off, oil prices rose in negotiated stages toward international levels (but never reached them) and both sides backed off their aggressive policies.

The federal, Alberta and Ontario governments co-operated to buy interests in Syncrude Canada Ltd.'s oil sands project, which was driven to the brink of collapse by the withdrawal of a corporate partner. The Alberta and Saskatchewan governments, alarmed by a high-profile movement of drilling rigs to the United States, adjusted taxes, lowered royalty rates and developed exploration incentives. The federal government partially restored the deductibility of royalties by introducing a new system called the "resource allowance."

The spell of harmony soon ended because the basic issues continued – whether to let Canadian oil prices rise to match international levels, then how to distribute the resulting new revenues as corporate profits, provincial royalties and federal taxes. The Canadian bill for oil imports mushroomed from $945 million in 1972 to $4.5 billion in 1979. While exports helped to offset this expense in Canada's balance-of-payments ledger, the net surplus in its oil and gas entry fell to $15 million in 1980 compared to $1.3 billion in 1974.

Although Canada's regulated wellhead oil price rose in mid-1979, the new level of $86.49 remained only 60 percent of the international level. The gap became a chasm when revolution in Iran drove world prices to nearly $270 per cubic metre.

THE NATIONAL ENERGY PROGRAM

This was the source of the headaches – in every field from economic planning to federal-provincial relations – that confronted Joe Clark every day of the nine-month life span of his minority government. Trudeau, called back from an announced retirement after the Tories' unexpectedly rapid fall, resurrected the central themes of 1970s Liberal energy policy as he led his party to a majority victory at the polls. After a brief round of unsuccessful negotiations on prices and revenue-sharing with Alberta, all the traditional federal themes were assembled in one package with sharp new teeth, the National Energy Program, announced in the budget of October 28, 1980.

Federal claims to a greater share of oil revenues were enacted as a renamed version of Clark's "self-sufficiency tax." On the other side, the NEP aimed to kill two birds with one stone – developing supplies and strengthening Canadian-owned companies – with PIP. This Petroleum Incentives Program paid up to 80 percent of exploration expenses, but the generosity was reserved for Canadian-owned companies working on federally-controlled Arctic and offshore territory. Ottawa also adopted a "back-in" right to take a 25 percent interest in any new production on these frontiers. The 1970s system of price regulation and export taxation continued.

Terra Nova well testing on the Grand Banks of Newfoundland, 1988
(Petro-Canada)

Led by an irate Lougheed, who compared the NEP to a rude invader blundering into Albertans' living rooms, the Western provinces mounted furious protests. Alberta adopted a version of OPEC's embargoes, by cutting the province's production 15 percent. Plans were announced to withhold approval from new oil sands projects. Court actions were launched.

The racket took a year to die down. In the fall of 1981, the provinces and Ottawa agreed to a formula that would let domestic oil prices reach 75 percent of international levels, with gas trailing along. Alberta restored oil production. Provincial versions of PIP were devised. Ottawa took some of the glow off the federally-controlled drilling frontiers by phasing out tax allowances for exploration expenses that were so generous they were called "super-depletion."

The revival of co-operation spread to the East Coast by 1982, when the federal and Nova Scotia governments made an agreement on offshore oil and gas. Goaded by visions of PIP-inspired discoveries, Atlantic Canada rapidly developed Alberta-like ambitions. They set aside issues of resource ownership to concentrate on revenue-sharing, with Nova Scotia allowed to keep all petroleum revenues until it ceased being a "have-not" province with income below the national average. A similar agreement took three more years to reach with Newfoundland, which sought outright ownership of discoveries like Hibernia far out to sea on the Grand Banks and only relented when the claims were defeated in the Supreme Court of Canada.

INTERNATIONAL CIRCUMSTANCES FOILED CANADIAN PLANS

At the same time, the NEP started coming apart. Once again, international events launched a tidal wave of changes in Canadian energy policy. The high prices triggered by the Iranian revolution turned out to be only temporary spikes on international commodity markets driven by a volatile mixture of speculation and fear of wars, supply disruptions or both in the Middle East. Like Clark's ill-fated proposals, the NEP rested on projections of price trends far into the future. A decade of OPEC price increases had stimulated new exploration and development outside the cartel, especially in the North Sea. Combined with conservation, substitution and demand shrinkage brought about by the 1970s price spiral, the new supplies forced OPEC to reverse directions. The cartel cut its benchmark price by 15 percent to $29 US per barrel. Agreed Canadian prices were cancelled in mid-1983, and a previously long line-up of multibillion-dollar oil sands projects began to shrink as their sponsoring consortia revised their expectations for oil prices.

OPEN FOR BUSINESS

At the same time, Canada's political complexion changed. Trudeau carried out his delayed retirement as Liberal leader. Successor John Turner proved to be unequal to the task of restructuring a tired administration. Brian Mulroney, a Quebec lawyer who took over from Clark after dissidents forced him to resign his leadership following the 1979-80 debacle, piloted the Tories to a majority in Parliament in a 1984 election.

Once again, international circumstances placed energy issues on the top of a new government's policy agenda. But Mulroney, who had also courted Western voters by promising a change, inherited an entirely different atmosphere. By the time he took over as prime minister, oil was abundant on international markets to the point where the price was falling below the lowered OPEC official postings. The same held true for natural gas. Among the last actions of the dying Liberal government was to inaugurate a gradual liberalization of border price controls after new supplies surfaced in the United States and its buyers developed resistance to high Canadian postings. Forecasts of food as well as energy shortages by agencies like The Club of Rome had proved to be false. The United States had safely dismantled much of its 1970s regulatory apparatus, and the immense personal popularity Ronald Reagan enjoyed for much of his presidency rubbed off on his advisers' new economic doctrines of "decontrol."

The new ideas about economics had taken hold so well across Canada that the Tories faced little protest as their energy minister, Vancouver MP Pat Carney, proceeded rapidly to dismantle the NEP. In March, 1985, the federal, Alberta, B.C. and Saskatchewan governments signed a Western Accord on how to go about the change.

Virtually all special oil regulation died at once, a schedule was set to phase out PGRT and PIP and a longer decontrol process was launched for the more complicated gas side of the industry. Within hours of the accord's announcement, Canadian oil prices were being set freely as refiner postings calculated purely on the basis of business considerations: international market movements, currency exchange rates, and adjustments for transportation costs and product quotas.

The Conservatives defanged foreign-ownership watchdog FIRA and changed the name to Investment Canada. Ottawa also began a long series of "privatization" sales of industrial investments, led by a gradual dismantling of Canada Development Corp.

The Tories' open-for-business message was underlined by the appointment of Roland Priddle as chairman of the NEB in 1986. A Scottish-born economist, Priddle worked for the international Royal Dutch Shell organization before Ottawa recruited him to help implement the pro-development NOP. He kept a low profile during the NEP and played a leading role in crafting the Western Accord. He inaugurated a version of energy policy that was dubbed the Priddle Principle by observers who welcomed him as a breath of fresh air in the Ottawa that wrote the book-length NEP. In a speech applauded enthusiastically by Calgary industry leaders, he said history had proved that the life expectancy of an energy policy stands in inverse relationship to its length and complexity.

Chapter 13

The Industry and the Environment

Rangeland sweeper restoring ground cover after pipeline installation, 1991
(NOVA Corporation, 318033)

COUNTING THE ENVIRONMENTAL COSTS

If there were any doubters left, the Canadian Bankers Association (CBA) served notice in late 1991 that anxiety over keeping the land, air and water clean had become a central and practical concern for all involved in industry. The notice came in an explanation for growing insistence on "environmental audits" of assets involved in sales of projects, and refusals to finance transactions or developments with loans unless potential pollution hazards uncovered by the investigation were cleaned up. Driven by public demand for ever-higher standards, a combination of government regulation and court rulings forced the financial community to accept responsibility for policing the environment. It was a matter of self-defense. Lenders stood to lose to the extent that borrowers were hurt by pollution damages or clean-up costs. Seen through green filters in a CBA pamphlet entitled "Sustainable Capital," oil-and-gas facilities stood out as cause for concern to the point where prospects were raised for a fundamental shift of capital away from resource extraction.[1]

The bankers warned that:

> *The aggregate effect of lenders' reluctance to lend to environmentally-risky businesses can be expected to alter the composition of credit markets in Canada. At the end of 1990, for instance, banks had $2.1 billion in non-mortgage loans outstanding to the oil and gas industry in Canada, $775 million outstanding to the mining industry and $540 million to the chemical industry.*

Impressive as that amount of lending business might sound, the banks pointed out they had much wider scope to earn interest by placing funds elsewhere:

> *At the end of 1990, Canadian banks had $95 billion in non-mortgage loans outstanding to private Canadian businesses. This suggests it would be possible for lenders to reduce lending commitments in Canadian credit markets where they face serious environmental risks while still maintaining large overall lending portfolios.*

This transfer of petroleum discoveries and facilities over to the liability side of bankers' ledgers, as environmental risks instead of the sweet smell of success at finding resource wealth, capped a long process of evolution.

EARLY ECONOMIC CONSERVATION MEASURES

As in other fields ranging from farming to wildlife management, economic conservation has been a central theme in most of the oil industry's history. Public and government ambitions to husband resources, minimize waste and achieve the greatest possible production spawned key institutions such as Alberta's Energy Resources Conservation Board and the National Energy Board.

Anxiety over the effects of resource development on the quality of the land, air and water came much later. Economic depression, war, faith that industrial growth could generate ever-rising living standards and shortages of established science on complex ecosystems, long drowned out fears of pollution. But from its birth, this concern demonstrated its potency. A dramatic turning point came in December of 1952, when an atmospheric inversion led to an environmental tragedy in London, England, on a scale to draw global attention. A severe accumulation of smog was blamed for as many as 4 000 deaths and brought the city virtually to a standstill. The British government blazed a trail for other nations by passing a Clean Air Act and allied environmental legislation.

# THE INDUSTRY AND THE ENVIRONMENT

THE PUBLIC AWOKE IN THE 1960s

By the 1960s, a general sense of a need to preserve environmental quality became widespread. Environmental concerns spread out from small interest groups. The thinking evolved from a limited idea of preservation, such as protecting endangered species and scenery, into a global anxiety that technology threatened an entire planet that came to be seen as a single living ecosystem where all the parts and their health are interconnected.

Canada followed international landmarks in the development of environmental consciousness. In 1962, a ground-breaking biography of the earth as a troubled planet Rachel Carson's *Silent Spring* stood out as a global best-seller. In 1968, the Club of Rome formed as an international association of thinkers aiming to think about industry and the environment on a global scale. The group's first effort, called Project on the Predicament of Mankind, concluded that the planet could not support projected rates of growth by population and industry beyond the year 2100. Published in 1972 as an international bestseller entitled *Limits to Growth*, the book predicted the environment would soon be stressed to a breaking point by increasing pollution, crowding and resource extraction. World energy supplies, especially non-renewable resources led by oil, were described as headed swiftly toward extinction.

Canada's petroleum industry first felt the effects of this changing public consciousness on its geographical frontiers. Fledgling exploration efforts in the Arctic and offshore of the East Coast drew wide attention and became targets for increasingly close scrutiny. Effects of oil spills elsewhere in the world were matters of high-profile public record. Questions naturally arose about the potential for damage in the Beaufort Sea or on the Grand Banks of Newfoundland. An accident in Canadian waters in 1970 the grounding of the tanker *Arrow*, which spilled its cargo into a lobster fishery off Nova Scotia – led environmentalists to suggest potential for an environmental disaster would inevitably accompany any offshore production developments.

COMMON LAW HOLDS POLLUTERS RESPONSIBLE

The new environmental consciousness also spread rapidly in the industry's established producing areas. Basic ingredients had been there longer than the oil industry. The federal government created Canada's first national park in Alberta, at Banff in 1885. By the time the oil industry reached Alberta, as an heir to nineteenth century common-law tradition, the province stood armed with the basics needed to deal with pollution. The British courts, which long remained the final authority for their Canadian counterparts, had established the idea of liability for public hazards, including pollution.

Anyone harboring a hazardous substance and allowing it to escape to hurt people or property was held to be liable for the damages.

This principle served as a foundation for much of the earliest environmental regulation of petroleum development. For example, the industry produces nearly as much salt water as oil, posing a potential danger of turning farm land into salt marshes and fouling drinking supplies. Early on the industry applied environmentally-safe practices for disposing of this water, such as drilling special wells to return it deep underground.

CLEANING UP "SOUR" GAS

While most environmental protection practices owed their development to quiet co-operation between industry and regulators, some practices resulted from public indignation. The first public action to counter pollution in Canada came when early natural-gas producers in southern Ontario sold quantities of "sour" production, laced with smelly and hazardous hydrogen sulphide. Consumers complained until Union Gas Co. built a processing plant in 1924 to "sweeten" the production by extracting the hydrogen sulphide. That same year, Royalite Oil built a similar plant in Turner Valley to sweeten production from the Royalite #4 well.

Although sour gas's obnoxious rotten-egg smell ignited the early environmental initiatives, the public soon began to understand that hydrogen sulphide was also toxic. Brief surges to high concentrations first deaden the sense of smell, then paralyze the lungs, cause unconsciousness and bring on death unless help arrives immediately.

Turner Valley residents developed an industrial version of "street-proofing" against the substance. It was not foolproof and led many to take risks that would be considered unacceptable today, such as tapping into pipelines carrying unprocessed sour gas to obtain household fuel supplies. Use of this gas was reasonably safe so long as a pilot light kept it burning, but became an immediate danger if the flame was blown out. It was standard practise for families to run out of their homes if pilots lights blew out, in a race to beat the buildup of lethal quantities of sour gas.

Practical economics – the discovery of sulphur production as a business proposition – came to the rescue. In 1952, sulphur-recovery equipment was installed at the Madison processing facility in Turner Valley and at Shell Canada's Jumping Pound plant, west of Calgary. The installations converted hydrogen sulphide to a yellow mineral with some value, while byproduct steam and less offensive sulphur dioxide were vented to the air.

Turner Valley residents tended to tolerate sour-gas problems because their livelihoods relied on petroleum production. When the industry spread rapidly in the 1950s after the Leduc discovery, new neighbors were not always so accommodating. Unfamiliarity with oil-and-gas operations helped create fear. Industry leaders and regulators acknowledged that public scepticism had some foundation during the early days, because knowledge of hazardous gases and technology for processing at that time was basic and unsophisticated.

PROVINCIAL STANDARDS ESTABLISHED NEW GUIDELINES

Nervousness about industry operations at Nevis and Pincher Creek led in 1957 to Alberta's first attempts to measure emissions from gas-processing plants. British American (BA) installed candle stations, or "bird cages" akin to safety devices used in methane-prone coal mines, in its Pincher Creek plant to measure hydrogen sulphide and sulphur conditions in the nearby air. In 1958, the company installed a more advanced logging device to measure emissions continuously. BA employed similar devices at its Nevis plant. Public complaints that year led the Alberta Department of Health to study the design of waste-gas incinerator stacks, then to insist that BA more than triple the height of one it was building.

The province lived up to its responsibilities under the Canadian constitution, as well as the British common-law, by pioneering environmental protection. Like natural-resource ownership, the field is under provincial jurisdiction unless environmental matters take on a national flavor through transportation, dispersion, trade or criminal nuisance.

Largely due to long experience with gas-plant emissions, Alberta became the first Canadian province to exhibit the enhanced environmental awareness that became a hallmark of the 1960s. Alberta's Health Department opened the decade by setting up an Air Pollution Control Branch. The province enacted a new set of regulations that covered all industries, but monitored compliance by sour-gas processors most carefully. The department held seminars to teach industries how to meet the new pollution-control standards. Also in 1960, the new branch developed an air-quality monitoring network around sour-gas sites. Responsibility for the monitoring gradually shifted to plant operators. To receive provincial approval to build new plants, operators also had to follow guidelines that put increasing emphasis on water-handling facilities, and on generally raising the quality of waste water.

Alberta's Energy Resources Conservation Board (ERCB) played a direct, pioneering role in pollution control. Although at some points in the industry's development taller stacks for dispersing pollutants was the only technologically practical method of control, the ERCB consistently maintained there would have to be a better answer. As years went by and the board continually advocated higher levels of recovery of sulphur and other byproducts from gas, the ERCB maintained that environmental considerations should help drive this basic resource-conservation policy. As required percentages of sulphur recovery from raw natural gas rose, volumes of hydrogen sulphide and sulphur dioxide emitted into the atmosphere automatically dropped.

In the 1970s, the early trends of the fifties and sixties intensified, to spawn most of the legislation that currently governs environmental matters in Canada. The seventies also saw the federal government become increasingly active, both directly on frontier territory where the environment is under its jurisdiction and indirectly in the provinces by proposing uniform national standards for the industry.

Once again, the Alberta home of the petroleum industry was quick off the mark. In 1971, one of the last acts of the aging Social Credit government created the provincial government's Department of the Environment. The Socreds also enacted a Clean Air Act. Within a year, their Conservative successors raised sulphur-recovery standards for gas processing into the high end of the 90 percent range. In the 1950s, the rules allowed recovery rates in the low 80 percent range. Beginning in 1971, Alberta also required gas-plant operators to monitor sulphur dioxide emissions continuously and report the results to the environment department. Computer-aided improvements in plant control, allowing feedback from emission monitors in the plants, greatly helped efforts to meet the rising emission standards. The industry also began installing second-stage sulphur-recovery equipment to clean up what had previously been considered waste gas.

In Ottawa, the federal Department of the Environment established national air-quality objectives in 1971, to set control levels for a series of compounds including sulphur dioxide. Only Alberta and Saskatchewan adopted the maximum standards suggested by the federal guidelines. Federal environmental legislation in the 1970s included the Arctic Waters Pollution Prevention Act and an Oil and Gas Production and Conservation Act, which extended federal pollution-control jurisdiction to Canada's undersea continental shelves. New federal regulations also were introduced to control refinery pollution, requiring new and higher rates for sulphides.

JUSTICE THOMAS BERGER AND THE MACKENZIE VALLEY PIPELINE COMMISSION

The industry gave a dramatic demonstration of the extensive environmental controls and requirements that had evolved by 1974 with proposals to build an Arctic natural-gas pipeline. Canadian authorities wasted no time in deciding too little had been done even though the project's sponsors submitted a two-volume environmental statement and 16 volumes of research data to the National Energy Board and the federal Department of Indian and Northern Affairs. The Liberal minister responsible, Jean Chretien, responded by appointing British Columbian Justice Thomas Berger to run a Mackenzie Valley Pipeline Commission.

After months of high-profile hearings in the north and elsewhere, the commission recommended a 10-year moratorium on industry along the Mackenzie River to let the region and its residents prepare for development. The NEB enacted the recommendations by choosing a route along the Alaska and Dempster highways over the initial Mackenzie Valley proposal by Canadian Arctic Gas. Even though glutted markets, falling gas prices and

THE INDUSTRY AND THE ENVIRONMENT 149

high construction costs eventually shelved the Arctic project, the industry inherited a message never to be forgotten. Berger effectively served notice that in future, the environmental, economic and social effects of petroleum projects would be major considerations in any decisions to approve them.

INDUSTRY RESPONDS WITH ENVIRONMENTAL INSTITUTIONS

The trend was well-established on other fronts. In 1974, Chretien's department also told the industry that drilling would only be allowed in the Beaufort Sea once environmental studies were completed. Besides studies by particular companies to support their own projects, the growing emphasis on environmental issues spawned industry co-operatives and long-lived committees of business and governmental officials to work on safeguards.

This collaboration sired an alphabet soup of technical bodies. An early example was APIGEC, the Alberta Petroleum Industry-Government Environmental Committee, which tapped business expertise to help resolve a variety of concerns. The Arrow tanker accident offshore of Nova Scotia served as a catalyst for the formation of a network of oil-spill co-ops across Canada. These evolved into PROSCARAC, or the Prairie Regional Oil Spill Containment and Recovery Advisory Committee. By the 1990s, PROSCARAC became a permanent institution incorporated by participants from throughout the industry.

The co-op approach stems from a conviction that since accidents are uncommon, yet affect the image and regulation of the entire industry, companies should pool their expertise and equipment to contain and recover spills in an organization akin to a civic fire department. As an economic benefit, oil producers spare the cost of buying and hiring all their own cleanup equipment by becoming able to draw on the pooled talent.

PITS, the Petroleum Industry Training Service, also dates from the birth of the present-day environmental movement. PITS serves as an oil field staff college with a curriculum that since the 1970s has included instruction on environmental matters ranging from emergency planning to detecting and caring for archaeological, paleontological and historical resources. At the same time, at the highest level of industry management and planning, trade associations and the senior companies involved in oil sands projects have built up libraries of environmental information, partly by financing research projects of specialty consultants and academic experts. The effort has ranged from quiet technical studies to high-profile, big-budget task forces such as an $11-million acid rain research program carried out in the mid-1980s by the Canadian Petroleum Association, the Alberta government and the ERCB.

Environmental projects likewise proliferated on the federally-controlled geographic frontiers of the Canadian industry. AMOP, or the Arctic Marine Oilspill Program, was set up in 1977 by Environment Canada and drew on industry research. COOSRA, the Canadian Offshore Oil Spill Research Association, helped feed AMOP with a multimillion-dollar set of 26 techni-

Archaeological excavation preparing for pipeline construction,
Crowsnest Pass, Alberta, 1993
(NOVA Corporation, 337520)

cal investigations. Most offshore oil operators belong to a seagoing version of PROSCARAC called ESRI, for East Coast Spill Response Inc. It has carried out trial runs at cleaning up mock spills in tandem with the Canadian and American coast guards and the Canadian Armed Forces.

SUSTAINABLE DEVELOPMENT

Although predictions about the imminent exhaustion of resources and the environment by global trend-setters like the Club of Rome did not come true, they ignited a transformation of industrial development. The power of the change showed when a 14-company drilling consortium led by Shell Canada Ltd. in 1986 made a major discovery of sour gas, which was also rich in liquid-fuel by-products.

The discovery was only a larger version of decades of gas finds and production projects in a prolific region 120 kilometres northwest of Calgary between the village of Caroline and the town of Sundre. Yet this time, going from discovery to ERCB approval for construction of a plant-and-pipeline network took four years of wrangling with second-biggest owner Husky Oil Ltd., negotiation with the communities and board hearings. Issues went far beyond technical safety matters like the durability and length of sour-gas pipelines, to community concerns such as selecting the processing-plant site, whether a new one should be built at all, effects on farm soils, the distribution of employment benefits, and controlling noise, traffic and the labor force during construction and afterwards.

In hindsight, Shell president Jack MacLeod described the often painful process as an early practical exercise in applying a doctrine for living together that environmentalists and industrialists evolved, called "sustainable development." In an address to an industry audience, MacLeod warned that the movement for business accountability had reached the point where companies stood to lose their legitimacy among customers, owners, employees and governments alike if they failed to honor the new social and environmental contract.

"In fact, the corporation has no choice," MacLeod said. "The corporation must ensure all its operations are environmentally sustainable or suffer loss." He outlined an agenda that ranged from tightening controls on production operations to conserving electricity and thereby reducing demands on carbon dioxide emitting, coal-fired power plants, by switching off every third fluorescent light tube in Shell's corporate headquarters. He invited his peers in the industry to take sustainable development personally, for their own good in the long run. MacLeod, a 38-year veteran of the petroleum industry, called himself "a sustainable-development freak." He was not kidding. He devoted his last months before retirement to the concept, leading long meetings about it among Shell staff as well as serving as chairman of Learning for a Sustainable Future, a national committee of industrialists working on a new education program for Canadian schools.

Prodded by environmentalists, the public, regulators and pride in their industry's history of self-starting technical innovation, MacLeod's peers took his advice. A new industry task force on plant decommissioning launched development guidelines for an estimated $5 billion worth of clean-up and abandonment jobs forecast at 200 000 small oil-and-gas facilities approaching retirement age across Western Canada.

Clipping willows for replanting after pipeline construction, 1993
(NOVA Corporation, 337502)

This voluntary commitment reflected widespread awareness in the industry that it could count on permanent pressure to adopt ever-higher standards. The need to respond to environmental concerns attained the level of a life-and-death matter for petroleum companies when high-profile accidents in transportation and exploration coincided with the greening of the public consciousness in the 1980s. The mishaps reinforced the sensitivity with an idea that saving the water, air and land had become practical issues.

The global oil transportation network held centre stage for months after Exxon Corp.'s tanker *Valdez* ran aground in March of 1989 and spilled more than 300 000 barrels on pristine Alaskan waters and coastlines. The tragedy triggered inquiries around the world. All asked the same questions: Can it happen here and how could it be prevented? Canada was no exception. Prime Minister Brian Mulroney appointed an inquiry tribunal under a prominent Vancouver lawyer, David Brander-Smith, to hold hearings across the country as Ottawa's Public Review Panel on Tanker Safety and Marine Spills Response Capability.

When the report came 18 months later, in November of 1990, the tribunal said tragedy not only could happen in Canadian waters — early warning signs showed shortly before the Valdez accident. In the opening lines of a 263-page catalogue of hazards and actions to reduce them, the panel described a collision close to Canadians' homes. Although the brush between a tugboat and a tank barge spilled less than 7 000 barrels of oil offshore of Washington state, it was enough to be noticed along the coast of British Columbia too. "A spill of the same magnitude as that from the Exxon Valdez could happen at any moment in Canadian waters. Indeed, without better prevention efforts, it will happen," the panel predicted. It pointed out that the stakes included the cleanliness of the Great Lakes and the Arctic, as well as the Pacific and Atlantic coasts.

By mid-1991, the federal transportation, environment, fisheries and oceans departments adopted prevention policies in co-operation with the industry group most directly affected, the Canadian Petroleum Products Institute (CPPI), an alliance of "downstream" refiners and marketers. The policy stopped short of enacting Brander-Smith's call for construction of a new generation of safer, double-hulled tankers (to be financed by a tax of $2 per ton on water-borne oil shipments or half a cent per litre by the time the levy was passed on to consumers at the gas pumps). But the government and the CPPI adopted a battery of changes such as $1-million fines for polluters, a $40-million upgrading of the industry's marine spill-response network, increased inspections of shipping and technical refinements of navigation procedures.

Inland, in the exploration and production industry's home base, the industry contended with an Alberta version of a tanker wreck – a sour-gas blowout in December of 1982. AMOCO Canada Petroleum Co.'s runaway Lodgepole well, southwest of Edmonton, killed two blowout-control specialists from Texas, sent 16 people to hospital and spewed out enough rotten-egg smelling hydrogen sulphide to annoy sensitive noses as far away as Winnipeg. Taming the high-pressure blowout took 68 days. ERCB inquiries and reports kept the case before the public for another two years. The incident spawned a new generation of safety regulations, requiring the industry to designate hazardous drilling targets as "critical wells" and to accompany the drilling with elaborate safety precautions.

Ten years later, sour gas regulation was still evolving towards ever tighter forms. Canadian Occidental Petroleum Co. set off a major public inquiry by reviving dormant proposals to drill for sour gas on the eastern fringes of Calgary. Following the pattern set by Caroline, the Calgary case rapidly developed into a marathon airing of public fears, then a general review of safety and environmental risks and safeguards. Prodded by Calgary health authorities and anxious community leaders, the ERCB suspended its review of Canadian Oxy's drilling application in order to review accepted standards for "setbacks" or the distances between gas production facilities and other land uses. The inquiry rapidly drew in the entire industry, with Canadian Oxy warning that tough new restrictions sought by nervous Calgarians "would needlessly sterilize a large percentage of the province's immense sour-gas reserves, perpetually denying Albertans the enormous economic benefits of their development."

A year after that warning by Canadian Occidental the case was still dragging on with no quick resolution in sight. It underlined the strength of environmental consciousness. The industry was unable to speed matters up even when it reminded the public and the ERCB that sour gas represented a major branch of development that earned $2 billion a year in revenues and provided thousands of jobs.

Chapter 14

The Impact of the Petroleum Industry on Canada

Rush to purchase oil stock in Calgary the day after the discovery of oil in Turner Valley, May 14, 1914
(Glenbow Archives/NA-601-1)

BUST AND BOOM – THE UNPREDICTABILITY OF THE PRICE OF OIL

When they created the Western Accord to dismantle the National Energy Program, federal, Albertan, Saskatchewan, British Columbian and industry leaders predicted it would unleash an "engine of growth" for Canada's economy. That was in March of 1985. Seven years later, Calgary oil field service contractor Rod Cleveland drew no quarrels by declaring "it's terrible here." Instead, he attracted envious offers from other specialists to enlist in his camp as he led a five-company caravan halfway around the globe to work among camels on desert wells in an oil field called Uzen beside the Caspian Sea in Kazakhstan. "This is a godsend," said partner Gordon Mitchell.

The departing oil field veterans were in good company. Their joint venture with a production unit of the southeastern state of the dismembered USSR, CanaKaz Global Oils Inc., expressed the strategy of the Canadian industry for at least the mid-1990s. Go international. Especially to Russia, if possible. Or at least to the former possessions, like Kazakhstan, that formed the Commonwealth of Independent States after the demise of Communism and the Soviet empire. CanaKaz only joined an exodus that already included a cross section of the Canadian industry, ranging from junior firms like Wega-D Geophysical Ltd. and Hurricane Hydrocarbons Inc. to traditional mainstays such as Gulf Canada Resources Ltd., PanCanadian Petroleum Ltd., Norcen Energy Resources Ltd. and Canadian Fracmaster Ltd.

The economic turning point was made official during a ceremony in Calgary on September 3, 1992, to mark the 25th anniversary of the Geological Survey of Canada's oil and gas arm, the Institute of Sedimentary and Petroleum Geology (ISPG). Retired ISPG director Walter Nassichuk declared, "Siberia is the next frontier for the Canadian industry." He threw in an offer to help the exodus, with a reminder that the agency had plenty of useful advice to give as a result of participating in 12 technical projects in the new promised land.

At the same time, federal deputy energy minister Bruce Howe consigned the engine-of-growth idea to the realm of dreams best passed over in silence. He made no apologies for the accord. He gave no hint that any changes would or could be made in the national hands-off policy, reinforced by free trade commitments to the United States, of letting market forces determine prices, drilling and development of Canadian oil and gas reserves. Instead, with a blend of personal sympathy but unyielding commitment to the policy, he put in a nutshell the history propelling entrepreneurs like Cleveland.

Howe said:

> *"The fundamental issue in the industry in Canada and the rest of the world is the price of oil. There is nothing any government can or will do about the price of oil." Howe acknowledged, "The economics of the industry are tough. The industry is solving them. It is very difficult, the restructuring, the reduction of costs. The restructuring is troublesome. But in the larger sweep of time, Canada has enough oil and gas. Some of it is expensive, in the Arctic and offshore. Some of it is tough to get at, in the Alberta oil sands. But hydrocarbon reserves for Canada are not an issue for the foreseeable future."*

In the eyes of the industry, Howe recited only facts of life accepted by all. No hostility towards the federal and provincial governments, or towards the oil and gas production companies, showed when Cleveland described his struggles to survive at home in Western Canada. Nor did accusations of broken promises fly when the Petroleum Services Association of Canada confirmed that the experiences driving Cleveland out were the rule rather than the exception. From a thriving, optimistic, 17-employee concern on the day the Western Accord was signed, his drilling-mud firm shrank within a year to a one-man ship able to hire temporary help only for short contracts won in hard rivalry with peers. The industry had become an engine of unemployment. PSAC president Jerry Thomson reported that after falling 40 percent to 18 000 jobs at the start of 1992 (compared to 30 000 in 1985) employment in oil field supplies, services and equipment manufacturing was headed down again by another one-third to a skeleton crew of 12 000 within six more lean months. Exploration and production companies fare little better. Since the onset in 1989 of the restructuring described by Howe, high-profile cost-cutting by the big-name oil companies alone eliminated 12 360 jobs by mid-1992. Those terminations came on top of an estimated 5 000 dismissals in the first waves of the industry's transformation in 1986. Also by 1992, companies bent on achieving efficiency had put up for sale an estimated $4 billion in oil field properties to shed all but their most profitable, "core" properties. Nothing was sacred. Cost-cutting spread into the public voices of the industry, as the Independent Petroleum Association and Canadian Petroleum Association voted on a plan to merge into a smaller combination called the Canadian Association of Petroleum Producers.

Howe drew no quarrels when he fingered oil prices as the cause of the anguish and added that government were powerless to do much about it. The same economic forces that undermined the NEP foiled the plans of the government and industry planners alike by accelerating to whirlwind speed within months of the Western Accord.

OPEC sowed the seeds of its own undoing in the 1970s. High world oil prices led to the development of competing supplies, especially in the North Sea. When a global surplus developed in the mid-1980s, the Thatcher government and the British industry made far-reaching decisions to accept whatever prices the world market would bear rather than accept repeated invitations by OPEC leaders to co-operate on shutting in excess capacity. By late 1985, oil was being traded by tanker-load like potatoes or cotton, on "paper-barrel" speculation led by Manhattan's New York Mercantile Exchange of NYMEX. The bottom fell out of the paper and wet-barrel markets alike when Saudi Arabia declared intentions to behave like the British and compete for sales volumes rather than continue to serve as the "swing" producer, shutting in productive capacity in efforts to prop up prices. By mid-1986, prices fell by two-thirds to lows under $10 US per barrel on some tanker loads and averaged $15.10 for the year, or barely half the 1985 level.

Natural gas fared no better as its North American market went through a more complex, slower but equally drastic evolution that was comparable to the end of the orderly world oil market. With price control abolished in both Canada and the United States and regulatory agencies accelerating sales competition by ordering pipelines to open to all comers, average field or wellhead prices fetched by Alberta gas plunged to $1.20 Canadian per thousand cubic feet for the first half of 1992 compared to nearly $3 the year before the Western Accord.

While natural gas dropped consistently in the absence of any powerful producer cartels, oil markets made occasional comebacks as OPEC struggled to restore order and crises periodically broke out in the Middle East. But annual average prices stayed well below 1985 levels even through the worst international supply scare, which was ignited by the 1990-91 international blockade, the war against Iraq over its occupation of Kuwait. After rallying to $24.46 US per barrel for 1990, international oil prices dropped back to $21.45 the next year and were slipping slightly in 1992.

The painful restructuring accelerated as virtually the entire industry learned from hard experience to turn its back on mid-1980s predictions that the trouble was temporary. Formerly fashionable forecasts of a recovery, always starting a year or two down the road, were dismissed as broken "hockey sticks," in a bit of jargon coined to describe the optimistic shapes of hopeful predictions.

THE WEST BEFORE LEDUC

The sense of loss went beyond economic anguish. The industry meant far more than dollars to Western Canadians, especially in its home base of Alberta. An Edmontonian, John Barr, captured what oil and gas prosperity had meant in a prophetic essay published before the price collapses.[1] He saw the problems coming early, from the vantage point of a job as a public affairs director for Syncrude Canada Ltd., which had been built on a foundation of expectations that conventional drilling and production would taper off as wells aged and their reserves ran dry.

Barr, who grew up in the years surrounding the 1947 Leduc discovery, recalled

> Prior to this time, Edmonton was a backwater. If you wanted to see a really big city, you drove all the way to the 'big apple,' Spokane. If you wanted to go out to dinner – and in those days, everybody went out for dinner, except for to their sister's place – you either went to the MacDonald Hotel or the local Chinese cafe.

The career outlook was equally barren:

> If you wanted to amount to virtually anything in virtually any field, whether business, academia or government, you had to move to Toronto, or Vancouver if you could not quite make it in Toronto. Premier Manning would be premier for the rest of the century, and Dr. Praisegood Barebones of the Edmonton policy morality squad ensured the peace, order and good government

of the city by periodically busting up card games in the back rooms of cafes.

The change after the Leduc find was as fast and complete as the crisis brought on by the oil-and-gas price crashes 40 years later. "Oil cracked this tight little world and let in dazzling rays of change," Barr recalled. Alberta communities suddenly filled up with

swaggering, superconfident Marlborough Men with Oklahoma or Texas accents who winked at the girls, drove big cars and came from a different world. The Edmonton Eskimos hired a coach who later returned to the United States and came to symbolize bit-time college football. His name was Darrel Royal, and he lived down the street from me. He was handsome, he had a beautiful blonde wife who used to be a cheerleader and he called his little girl Sugar. He symbolized the beginning of a different kind of Alberta.

AMERICANIZATION, PROSPERITY AND URBAN GROWTH

Attitudes and attachments formed that contributed to the emotional heat Albertans radiated as they cheered Peter Lougheed and reelected his Conservatives with huge legislative majorities to the 1970s energy wars with the federal government. The industry's founding boom after Leduc "was 'Americanization' and we loved every minute of it. America was big time. The big league. Suddenly we were somebody, we were somewhere, and we were getting a piece of the action. Deep down, no Albertan who grew up in the 1950s could ever be truly anti-American. The Americans introduced us to the big time. They made it possible to be first-class Canadian for the first time, instead of just the dumb hicks from the West."

Barr was just one among thousands who grew up into better jobs, in bigger cities spawned by a stronger economy than their parents ever dreamed of seeing in the Great Depression of the 1930s or the stagnation in the early 1940s. Eric Hanson, in a book titled *Dynamic Decade*, documented the transformation brought on by oil in the years 1946-56. In the previous 10 years, an estimated 80 000 people left Alberta to cut its population to 803 000 by the year before the Leduc discovery and the total was thought to be on its way down to 750 000. The oil bonanza reversed the tide. Between 1951 and 1961, Alberta's population grew by 41.8 percent compared to the Canadian national average 30.2 percent, which was in turn high by international standards. The trend continued from 1961-71, when Alberta's population rose 22.2 per cent while Saskatchewan stayed flat and Manitoba grew only 7.2 percent. By 1981, Alberta had twice as many people as either Saskatchewan or Manitoba. In 1947, the three provinces had about the same population.

OIL AS A POWERFUL INVESTMENT MAGNET

Calgary and Edmonton became the fastest-growing cities in Canada. Calgary doubled between 1948 and 1958, then doubled again to 410 000 by 1971. Although this pace proved to be impossible to sustain even in the boom times before the effects of the NEP, national recession and double-digit interest rates took hold of the economy, those years set a record for annual increases when Calgary grew by 31 000 people in 1981 alone. In the 40 years after the Leduc discovery, the industry capital of Calgary grew to 636 000 from 100 000. As the governmental capital as well as headquarters of oil field operations and services, Edmonton saw its population multiply fivefold to 574 000 from 113 000. The growth spread far beyond the two principal cities. The oil sands capital, Fort McMurray, achieved staggering 25-fold growth to 30 000 between 1961 and 1981 as first the Suncor and then Syncrude plants were built. Communities across Alberta felt the expansion. Between the Leduc discovery and the onset of the oil and gas price drop in 1986, Lloydminster in the east grew to 17 356 from 2 541, central Red Deer expanded to 54 245 from 3 995 and Grande Prairie mushroomed in the northwest of the province to 24 471 from 2 267. The effects of oil spilled over into Saskatchewan, where communities near further discoveries like Weyburn and Estevan grew between one-third and 300 percent. In British Columbia, similar communities like Dawson Creek and Fort St. John achieved comparable expansion.

Imperial Oil's Leduc find turned out to be the key that opened up the entire Western Sedimentary Basin, stretching from eastern B.C. to western Manitoba and up and even into the Beaufort Sea. In quick succession, entrepreneurs translated the new geological understanding yielded by Leduc into major fields like Red Water, Golden Spike, Wizard Lake, Fenn-Big Valley and Bonnie Glen, each harboring more than 90 million barrels of oil reserves. Production soon saturated local markets. Alberta stopped importing gasoline from the United States. As Alberta crude began to move by railway tank cars to Saskatchewan and Manitoba, refineries there too became independent of imported oil. Rapid pipeline construction soon sent Alberta production to Central Canada, the middle-western United States and the northwestern states. Natural gas, always more complex to develop, followed by the late 1950s and early sixties. More abundant than oil, Western Canadian gas markets now span the continent from Long Island to the Los Angeles region thanks to steady additions to pipelines.

Until the 1986 oil and gas price failures, Alberta stood out as one of the most powerful investment magnets on the North American continent. Annual investment doubled in the 1950s and again in the 1960s. Between 1947 and 1986, the petroleum industry's exploration and capital expenditures in Alberta reached a total $63 billion. Even in 1986, the petroleum industry still stood out as its home province's principal money earner, with a 35 percent share of the Alberta Gross Domestic Product compared to three percent for agriculture. Before the Second World War, Alberta's income per capita was less than the national average. In an episode most would rather forget, the provincial government in the mid-1930s briefly imposed the only sales tax Alberta ever had in an effort to come back from defaulting on bond interest payments.

By 1966, the Albertan's incomes topped the national average and they stayed on the high end of the Canadian range throughout the 1980s. Even after taking over full responsibility for education, health and welfare and repaying most municipal government's debts in the 1970s, the provincial government had revenue surpluses to build up a multibillion-dollar Heritage Savings Trust Fund.

THE 1986 OIL PRICE COLLAPSE

Just as the public shared in the industry's good years, the hard times after the 1986 price collapse rippled into every corner of the province. Alberta's non-renewable resource revenues fell by more that half from $4.978 billion for the fiscal year ended March 31, 1985 to $2.132 billion for 1991-92. Although the province stopped short of kindling memories of the Dirty Thirties by refusing to impose a retail sales tax, it restored gasoline taxation abolished in the 1970s, raised income taxes, and restrained health, education and welfare spending – and still faced the prospect of continuing annual deficits and a rising deficit for the rest of the 1990s.

Like the deputy federal energy minister, no industry leaders were rash enough to predict when oil-and-gas prices would make a comeback to rekindle oil field activity and make the money flow in again. But also like Howe and the professional explorationists who observed the economic turning point during the 1992 anniversary of the Geological Survey of Canada, the industry pointed to an enduring legacy. Western Canada remains heir to a proud tradition of technical and business innovation that stands ready to revive development when energy economics warrant a rebirth, as well as to a storehouse of knowledge being turned to development of export markets by the geological, engineering, managerial, service, supply and manufacturing communities.

POTENTIAL DEVELOPMENT IN CANADA

The heritage spans the nation. Even in the industry's chaotic 1980s, energy provided an average seven percent of Canada's gross domestic product, 12 percent of export income and 17 per cent of investment outside of personal housing. Coal, nuclear power and hydro-electric dams made contributions. But the National Energy Board estimated that, as of 1990, oil, natural gas and its liquid by-products still accounted for nearly 70 percent of Canadian energy supplies. The Canadian economy ranked as the biggest user of energy on earth. Abundant supplies made possible the development of other industries that use energy heavily, such as pulp and paper, chemicals, mining and metal processing.

As Barr wrote on behalf of all who like him grew up and prospered in the 40 years between Leduc and the price collapses, "it has been a great ride."

Is that ride finished? By late 1992, after six years of unrelieved hard times following the 1980s collapses of oil and natural gas prices, the industry answered glumly in the affirmative. A federal government department, Employment Canada, recorded this verdict as virtually unanimous in a landmark report.[2] This bleak finding carried an unusually high level of conviction because the study was done by management and economics consulting firms rather than remote civil servants. The experts ran an 18-month investigation of the petroleum industry under the supervision of its trade associations.

The investigators recorded a startling reversal of fortunes: "The oil patch is no longer generally regarded as a good place to work. People who were once driven by challenge and opportunity are now driven by fear."

Even after counting rehirings by newcomer companies on the buying side of the asset-sales market plus long-term contract jobs, employment shrank by 14 percent or 10 200 positions between 1988 and the end of 1991. Hiring of university graduates in the science-based professions had all but halted, while purges of older staff eligible for early retirements left the industry with a work force averaging 38 years old. These remnants had no security: "The current outlook is worse than that which caused the previous downsizings in 1988-91 . . . about 60 percent of companies we interviewed felt they had too many staff and expected to downsize, many for the second or third time. So downsizing will be a fact of life at least until the mid-1990s."

With their anonymity protected by the inquiry's consulting firms, the major oil employers proclaimed their industry had outgrown its exploration phase, or youth, to enter a progressively leaner maturity. Even if oil prices bounced back to $32 a barrel and gas nearly doubled to $2.30 per thousand cubic feet by the year 2 000, the industry expected to enter the 21st century leaner than ever, with about 60 000 employees or six percent fewer than in 1992. If oil and gas prices stayed flat, no alternative was seen to a drastic, 32 percent shrinkage that would end about 21 000 more jobs.

This relentless cost-cutting grew out of the harshest lesson of the 1980s: Even if they stay essential to a modern economy, oil and gas remain businesses that must pay their own way. Legacies of inflated expectations, cluttered asset portfolios, fat payrolls and heavy debts left over after the optimistic 1970s brought the industry to its financial knees when deregulation, open markets and competition took over as economic driving forces from government controls and the Organization of Petroleum Exporting Countries. Rates of return in 1986-90 averaged 3.3 per cent on capital employed and 2.5 percent on equity – or less than half of both the cost of raising funds and performance by other Canadian industries even though they too were bogged down in recession.

Along with production technology, provincial royalties, municipal taxes and administrative procedures, the industry concentrated on changing its personnel as a cost it had at least some hope of managing efficiently while expecting global commodity prices to stay beyond anyone's control. The inquiry by Employment Canada ushered in a new phase of petroleum development by posing a fresh challenge to go with the traditional hurdles of geology, engineering and politics. "The industry has to resolve a human-resource planning dilemma," the report stated. The tall order for the 1990s became to revive morale and restore the industry's hopes to the point where it could continue to compete for Canada's best talent.

RIDING INTO THE SUNSET OR THE SUNRISE

As the 1992-93 hard winter of cold snaps and job terminations ended, the petroleum industry showed scattered but strong signs of warming up, along with the weather, across Canada. A new Alberta energy minister and the first woman to hold the post, Patricia Black, drew no quarrels for declaring, "It's a sunrise industry – it's not a sunset industry."

Similar encouraging words arrived from Bay Street, where Toronto investment analysts and securities houses likewise declared an impressive recovery to be developing. The financial houses said the oil companies' painful efficiency drive had paid off to make them the leading factor, during the last three months of 1992, in the strongest quarterly profit gain seen in more than two years among firms traded on the Toronto Stock Exchange.

Recovery showed across the trading board. Among the industry's top participants and most spectacular cases of "restructuring" headaches, Imperial Oil, Petro-Canada and Nova Corporation reversed 1991 losses to return to earning profits in 1992. At the smaller end of the industry scale, independent companies parleyed asset acquisitions from the giants into a 30 percent increase in production in 1992. And they made money at it, by being able to add the oil and gas output without also acquiring the overhead-cost problems that had plagued the big companies.

The outlook likewise brightened out in the oil fields and international markets for Canadian production. Ottawa's Department of Energy reported the industry in late 1992 reversed a late-1980s and early-1990s decline in oil production that had been triggered by the price collapse. Conventional drilling for refinery-ready crude accelerated under the stimulus of provincial royalty reductions. But the biggest gain was chalked up by the formerly hardest hit part of the industry. Production of abundant heavy crude reserves, a strategic endeavor to make up for natural depletion of light supplies, rose 10 percent into the 500 000 barrel-a-day range.

Participants in this specialty, like AMOCO Canada Petroleum Co., Imperial and Koch Oil Co. signalled that they intend to sustain the new highs by starting long-term field developments intended to attain large scales in the range of 40 000 to 50 000 barrels of heavy crude a day. Much of the credit goes to oil field counterparts of the reorganization of corporate head offices. Production costs steadily fell under a host of engineering improvements, led by horizontal drilling techniques that cut by as much as half the number of wells required for heavy-oil development. Also, a new pumping technique called "cold flow" allows some reserves to be tapped without costly injections of heat.

Natural gas turned in a stellar performance that showed signs of continuing, thanks to rising demand, new pipeline construction and successful adaptation to open, deregulated international markets. In the contract year that ended October 31, 1992, Canadian exports to the United States set a record that was double their best performances of the 1970s boom times by hitting two trillion cubic feet and a 10 per cent share of the total American gas market. Firming prices, coupled with favorable currency-exchange rates, generated a 23 percent jump in export revenues to $4.3 billion Canadian.

Further growth by the gas sector was virtually guaranteed, as construction proceeded on new Alberta-to-California pipeline facilities. Eastbound delivery routes also continued to add capacity. New projects, to be completed in the mid to late 1990s, cropped up to serve regions from Mexico to New England. Ottawa's energy department, in a forecast described as conservative because it refrained from speculating on the likelihood of success for new pipeline proposals, predicted substantial growth. By the 1996-97 contract year, exports were expected to rise another 25 percent to 2.53 trillion cubic feet.

Alberta's new energy minister, a veteran of 15 years in management positions in the industry – including the hard times in the mid-1980s – declared she had no doubts that the oil industry had far from finished a long history as a major engine of the economy. With her hopes fuelled by experience, Black declared Canadians can count on it: "This industry has been through hell and back. Because of the nature of its people, it'll survive and flourish. They're in for the long haul. You can kick them but you can't keep them down. They're risk takers and developers."

Notes to Chapter 1 – The Great Oil Age

1. Personal communication with Wayne Deschamps, General Manager, Safety Boss Ltd., September 10, 1992.
2. Stephen Leacock, "In Praise of Petroleum," *Imperial Oil Review* (August/September, 1930): 8-9. Leacock prepared his essay to commemorate Imperial Oil's 50th anniversary.
3. Victor Ross, *Petroleum in Canada* (Toronto: Southam Press Limited, 1917), p. 1.
4. Nordegg, Martin, *The Fuel Problem of Canada* (Toronto: The MacMillan Company of Canada Limited, 1930), p. 7.
5. Nordegg, p. 75-81.
6. Nordegg, p. 75.
7. Ross, Victor, *The Evolution of the Oil Industry*, (New York: Doubleday, Page and Company, 1920), p. 163-164.
8. Ernest C. Manning, speech on Canada-U.S. relations, delivered to the Senate of Canada on April 28, 1971. Reprint, p. 6.
9. Quaintly, the term used was "bounty."
10. Based on production estimates in Nordegg, p. 79 and p. 83. Nordegg's natural gas estimated for 1929 are anomalous. Compare his estimates to those in the Canadian Petroleum Association's *Statistical Handbook* for 1947, the first year for which consistent records are available.

	1929 (Nordegg)		1947 (CPA)	
	Production	Value	Production	Value
Central Canada	9.7 bcf	$5.3 m	8.3 bcf	$5.6 m
Alberta	19.0 bcf	$4.5 m	32.8 bcf	$2.2 m

11. William R.S. McLellan, "The Geological Survey of Canada and the Petroleum Industry: A Partnership of Discovery." Unpublished paper delivered at the Glenbow Museum Library, Calgary, September 30, 1992, p. 6.
12. Personal discussion with Robert Vallance, then public affairs manager for Gulf Oil Canada Limited, summer 1978.
13. These relative numbers applied in the early 1990s. Sources: Statistics Canada and Energy Mines and Resources.
14. National Energy Board, *Canadian Energy Supply and Demand, 1990-2010* (Calgary, Ministry of Supply and Services, June 1991), p. 194.
15. Ibid.
16. Ibid.

Notes to Chapter 2 – The Source

1. Robert H. Dott, et al., *Sourcebook for Petroleum Geology* (Tulsa: The American Association of Petroleum Geologists, 1969), p. 8.
2. Dott, p. 3.
3. *Canadian Petroleum Association Review* [Calgary], April, 1992.
4. R.M. Proctor, et al., *Oil and Natural Gas Resources of Canada*, (Geological Survey of Canada, Energy Mines and Resources Canada, 1983), p. 11.
5. Proctor.
6. Dott, p. 10.

7. Dott.
8. Max Ball, *This Fascinating Oil Business*, (Indianapolis: Bobbs-Merrill Company Inc., 1940), p. 304.
9. Ball, p. 295.
10. Dott, p. 11.

Notes to Chapter 3 – The Early Years

1. Many sources describe the inventions of the Ruffner brothers, but the most thorough is J.E. Brantley's *History of Oil Well Drilling*, (Houston: Gulf Publishing Co., 1971).
2. *Petrolia Discovery, The Discovery of Oil in Canada*, Petrolia, 1980, p. 1.
3. Brantley, p. 185.
4. Brantley, p. 206.
5. Brantley.
6. Kenneth E. Anderson, Bill D. Berger, *Modern Petroleum*, (Tulsa: PennWell Publishing Company, 1978), p. 8.
7. Anderson, p. 10.
8. D.W. Burns, "Cable-tool Drilling Method and Wooden Derricks Used," *Oilweek*, May 18, 1964, p. 62-63.
9. George de Mille, *Oil In Canada West, The Early Years*, (Calgary: George de Mille, 1970), p. 92.
10. de Mille, p. 100.

Notes to Chapter 4 – Oil Between the Wars

1. George de Mille, *Oil In Canada West, The Early Years*, (Calgary: George de Mille, 1970), p. 112.
2. de Mille, p. 117.
3. de Mille, p. 112.
4. de Mille, p. 140.
5. de Mille, p. 141.
6. Canadian Petroleum Association, *Canadian Oil and Gas: The First One Hundred Years*, (Canadian Petroleum Association, 1984), p. 4.
7. de Mille, p. 141.
8. Earle Gray, *Wildcatters*, (Toronto: McClelland and Stewart, 1982), p. 82.
9. A.W. Rasporich, "Oil and Gas in Western Canada, 1900-1980," *Canada's Visual History*, Vol. 70, National Museum of Man, (Ottawa: 1970), p. 7.
10. Aubrey Kerr, *Atlantic Leduc No. 3: 1948*, Calgary, S.A. Kerr, 1986, p. 31.
11. Philip Smith, *The Treasure Seekers, The Men Who Built Home Oil*, (Toronto: Macmillan of Canada, 1978), p. 96.
12. Anonymous, "Functions of Wartime Oils Not Properly Understood," *The Western Examiner*, Calgary, (May 1, 1943).
13. de Mille, p. 165.
14. de Mille, p. 187-190.
15. de Mille, p. 195.
16. Dome Petroleum, "Canol pipeline, A History of the Norman Wells to Whitehorse Oil Pipeline," *Beaufort*, I, IV (1984): 6.

17. *Beaufort*, p. 7.
18. Eric Hanson, *The Dynamic Decade*, (Toronto: McClelland and Stewart, 1958), p. 53.

Notes to Chapter 5 – The Leduc Years

1. Earle Gray, *Wildcatters*, (Toronto: McClelland and Stewart, 1982), p. 90-91.
2. Eric Hanson, *The Dynamic Decade*, (Toronto: McClelland and Stewart, 1958), p. 104.

Notes to Chapter 6 – The Pipeline Era

1. William Kilbourn, *Pipeline*, (Toronto: Clarke Irwin, 1970), p. 8.
2. Kilbourn, p. 9.
3. Hanson, p. 171-172.
4. Trans-Mountain Oil Pipeline Company, pamphlet for official pipeline opening, 1953.
5. Trans-Mountain pamphlet.
6. G. Bruce Doern and Glen Toner, *The Politics of Energy*, (Toronto: Methuen, 1985), p. 167-168.
7. Doern, p. 168.
8. Earle Gray, *The Great Canadian Oil Patch*, p. 133.
9. Kilbourn, p. 161.

Notes to Chapter 7 – Processing Gas

1. Canadian Petroleum Association, *Statistical Handbook*, (Canadian Petroleum Association), 1985, Section III, Table 11.
2. *Statistical Handbook*, Section III, Table 13.

Notes to Chapter 8 – Oil Sands and the Heavy Oil Belt

1. According to Alberta's Energy Resources Conservation Board, Canada's bitumen in place (virtually all of it in Alberta) is about 2.5 trillion barrels, which is more than double the world's entire proved oil reserves of about 1 trillion barrels. Underlying the oil sands is another vast resource, the carbonate triangle. This huge deposit of bitumen-saturated dolomite may contain another 800 billion barrels of bitumen.
The National Energy Board estimates total heavy oil in place in Saskatchewan and Alberta at about 37 billion barrels, 63 percent of which is in Saskatchewan.
While these raw numbers are large even by Middle East standards, they are somewhat misleading. According to the National Energy Board, the ultimately recoverable bitumen in the oil sands deposit is about 308 billion barrels; of heavy oil, about 6.7 billion barrels.
As the text explains, the amount of oil recoverable from these resources is heavily dependent on economics and technology. It is therefore possible that Canada will never produce these amounts.
For more information see:
National Energy Board, *Canadian Energy Supply and Demand, 1990-2010* (Minister of Supply and Services Canada, June 1991), p. 194.
World oil reserves from *BP Statistical Review of World Energy* (London, England: British Petroleum Company plc, June 1992), p. 2.

Resource in place in the carbonate triangle from a personal communication with Hans Maciej, then vice-president of the Canadian Petroleum Association, October 1989.

2. Robert E. McRory, *Energy Heritage: Oil Sands and Heavy Oils of Alberta*, (Edmonton: Alberta Energy and Natural Resources, 1982), p. 14.
3. McRory, p. 23.
4. de Mille, p. 40.
5. de Mille, p. 43.
6. J. Joseph Fitzgerald, *Black Gold with Grit*, (Sidney, B.C.: Gray's Publishing, 1978), p. 39.
7. Barry Glen Ferguson, *Athabasca Oil Sands: Northern Resource Exploration*, 1875-1951 (Alberta Culture/Canadian Plains Research Centre, 1985), p. 21.
8. "Synthetic" because the oil results from chemical alterations to the bitumen. The properties of synthetic oil are like those of a high grade conventional light oil. Syncrude's synthetic oil has a better carbon/hydrogen ration than conventional crude, and is thus a high-quality, premium crude oil.
9. Larry Pratt, *The Tar Sands: Syncrude and the Politics of Oil*, (Edmonton: Hurtig Publishers, 1976), p. 189-190.
10. Government of Alberta, *Proceedings, Athabasca Oil Sands Conference 1951*, (Edmonton, 1951), p. 175-176.
11. McRory, p. 32.
12. Darlene J. Comfort, *The Abasand Fiasco: The Rise and Fall of a Brave Oil Sands Extraction Plant*, (Edmonton: Friesen Printers, 1980), p. 105.
13. T.J. Latus, "Development in Western Canada in 1962," *Bulletin of the American Association of Geologists*, 47, 6, p. 1188.
14. Earle Gray, *The Impact of Oil*, p. 52.
15. Earle Gray, *Great Canadian Oil Patch*, p. 311.
16. David G. Wood, *The Lougheed Legacy*, (Toronto: Key Porter Books, 1985), p. 119.
17. Eric P. Newell, "The challenge of the world's largest storehouse of oil," *Canadian Speeches*, 5, 8 (December, 1991): 33-38.
18. Fitzgerald, p. 114.
19. Personal communication with former Alberta premier Ernest Manning, 1988.
20. Andrew Brown, "Shifting the Centre of Gravity," *The CPA Review*, IX, 2, (May 1985): 3.

The Wolf Lake formula, so named because it first applied to the Wolf Lake project, involved project-specific modifications to the federal government's National Energy Program. Announced in 1980, the NEP was eliminated in 1985.

By the Wolf Lake formula, the federal government would allow capital expenditures to be deductible in calculating the Petroleum and Gas Revenue Tax, and the project would be eligible for Petroleum Incentive Payments. In addition, the federal government would authorize oil exports from these projects if recommended by the National Energy Board.

For its part, Alberta would cut royalties to one percent for the first 18 months of a project. After that, gross royalties would increase by one percent every 18 months until reaching five percent. The rate would remain at five percent until discounted cumulative net revenue (gross revenue, less allowances for

capital expenditures, operating and material costs and royalties) reached payout. The subsequent royalty would be the greater of five percent of gross revenue or 30 percent of net revenue.
21. Peter McKenzie-Brown, "The Economic Benefits of Oil Sands Development," *The CPA Review*, IX, 2 (May, 1985): 10.
22. Hladun, Helene, "Shafts and Tunnels," *The CPA Review*, 14, 1, p. 3-5.
23. David Scollard, "Highlights from Husky's First 1/2 Century," *Horizons*, 6 (Husky Oil Operations Ltd., Calgary, 1987): 2-6.
24. Imperial, which already owned a refinery in Regina, bought out Maple Leaf Petroleum, in Coutts, Alberta. British American bought Northwest Stellarene in Coutts and Sterling Oil Refineries in Moose Jaw, Saskatchewan. For more information see: Brett Fairbairn, *Building a Dream: The Co-operative Retailing System in Western Canada, 1928-1988* (Saskatoon: Western Producer Prairie Books, 1989,) p. 67.
25. Fairbairn.
26. Anonymous, *50 Years of Refining at the Consumers' Co-operative Refineries Limited* (Regina: Consumers' Co-operative Refineries Limited, 1984).
27. Personal communication with R.D. Harris, Production and Supply Manager, Consumers' Co-operative Refineries Ltd., 1992.

Notes to Chapter 9 – Frontiers of Muskeg, Ice and Water

1. Max Foran, "Calgary, Calgarians and the Northern Movement of the Oil Frontier," *The Making of the Modern West: Western Canada Since 1945*, A.W. Rasporich, ed. (Calgary: University of Calgary Press, 1984), p. 117.
2. The Immerk well was dry, but Adgo F-28 was a multi-zone discovery of oil and gas.
3. Jim Lyon, *Dome – The Rise and Fall of the House That Jack Built*, (Macmillan of Canada, 1983), p. 119.
4. Information on Hillsborough No. 1 provided by Tom Cooney, Mobil Oil Canada, Halifax.
5. In practice, the question has not been formally resolved. As part of negotiations which led to the 1985 Atlantic Accord, the two governments agreed not to pursue the question any further in the courts.

Notes to Chapter 10 – From Crude to Refined

1. Ball, p. 312.
2. G.A. Purdy, *Petroleum: Prehistoric to Petrochemicals* (Vancouver, Toronto, Montreal: Copp Clark Publishing Company, 1957), p. 19-20.
3. J.T. Henry, *The Early and Later History of Petroleum*, p. 136.
4. Petroleum Resources Communications Foundation, *Our Petroleum Challenge: The New Era*, (Calgary, 1985), p. 15.
5. Imperial Oil Limited, *Third Submission to the Restrictive Trade Practices Commission on the State of Competition in the Canadian Petroleum Industry*, (Toronto: 1983), p. IV-2.
6. W.A.E. McBride, "Ontario: Early Pilot Plant for the Chemical Refining of Petroleum in North America," *Ontario History*, LXXIX, 3, (September 1987): 210-214.
7. Purdy, p. 28.
8. McBride, p. 210-14.

ENDNOTES for Pages 104–114

9. Imperial Oil, p. IV-2.
10. McBride, p. 217-219.
11. Kendall Beaton, *Enterprise in Oil, A History of Shell in the United States*, (New York: Appleton-Century-Crofts Inc., 1957), p. 570.
12. Beaton, p. 575.
13. Chapter 8 recounts the stories of two early refineries, the Consumers Cooperative Refinery in Regina and the Husky refinery in Lloydminster, which both contributed to heavy oil development.
14. Imperial Oil Limited, p. IV-3-4.
15. Texaco Canada Inc., *Submission to the Restrictive Trade Practices Commission on the State of Competition in the Canadian Petroleum Industry, The Refining and Marketing of Petroleum Products in Canada*, (Toronto, 1983, p. 86.
16. Imperial Oil Limited, p. IV-4.
17. James Lorimer, *Canada's Oil Monopoly, Highlights from the State of Competition in the Canadian Petroleum Industry*, James Lorimer & Company, Toronto, p. 35.
18. Energy, Mines and Resources Canada, Canadian Oil Markets & Trade Division, "Canadian Motor Gasoline Markets: 1980s, The Decade in Review," (Ottawa: December 1989), p. 5-6. This report notes that:
Average provincial taxes more than doubled to over 10 cents per litre in 1989. In Quebec and British Columbia road taxes tripled. Some provinces moved from ad valorem to fixed rates in an attempt to stem revenue losses from falling gasoline prices. Other provinces increased their ad valorem rates to make up the difference. In 1982 Saskatchewan dropped its tax on gasoline, re-introducing it in 1987, about the same time Alberta re-introduced its petroleum product tax.
19. Quoted in James G. Macgregor, *A History of Alberta* (Edmonton: Hurtig Publishers, 1981), p. 194.
20. Petro-Canada purchased the Gulf assets west of Quebec, while Ultramar purchased the Gulf assets east of Ontario.
21. Lorimer, James, ed. *Canada's Oil Monopoly: The Story of the $12 Billion Rip-off of Canadian Consumers* (Toronto: James Lorimer and Company, 1981), p. xx-xxi.
22. Ibid, p. 9.
23. "The Truth about Oil," *The Globe and Mail*, Wednesday, June 18, 1986. An editorial.
24. Restrictive Trade Practices Commission, *Competition in the Canadian Petroleum Industry* (Ottawa: Supply and Services Canada, 1986).
25. Restrictive Trade Practices Commission, Vol. I: "Introduction, Conclusions and Recommendations," p. 56.
26. The commission's actions were more a highly visible justification for the creation of the PMA than a motivator. The federal department of Energy Mines and Resources had been monitoring the industry's financial performance since 1977. However, the PMA had an additional mandate to investigate ownership and control trends, the flow of funds and research and development activities in the petroleum industry. Thus, it helped to measure the effectiveness of the National Energy Program's Canadianization initiatives, for example.
The PMA was dissolved in early 1992, although many of its functions are still carried out by Energy Mines and Resources.

Notes to Chapter 11 - Petrochemical Miracle Workers

1. Peter H. Spitz, *Petrochemicals: The Rise of an Industry* (John Wiley and Sons: New York, Chichester, Brisbane, Toronto, Singapore, 1988), p. 63.
2. Spitz, p. 65.
3. Vincent N. Hurd, "Petrochemicals," in James D. Hilborn, ed., *Dusters and Gushers: The Canadian Oil and Gas Industry* (Toronto: Pitt Publishing Company, 1968), p. 154.
4. A technical discussion of the difficulties of defining petrochemicals can be found in a book by the Chemical Division of Shell Oil Company of Canada, *The Canadian Petrochemical Industry* (Toronto: The Ryerson Press, 1956), p. viii. This volume, which is the source of most material for the early years of petrochemical development in Canada, maintains that "a petrochemical is a chemical compound or element derived mainly from petroleum or natural gas or their derived hydrocarbons and having applications in chemical processing."
5. The following table summarizes the distribution of large petrochemical plants in Canada, according to a survey in the August 17, 1992 issue of *Oilweek* magazine.

British Columbia:	4	Alberta:	19
Manitoba:	1	Ontario:	19
Quebec:	9	Nova Scotia:	1

However, a detailed aggregation by the Canadian Chemical Producers Association notes that in 1990 27 "establishments" in Alberta manufactured and sold $2.2 billion of petrochemicals; 32 in Quebec manufactured and sold $1.1 billion; and 88 in Ontario manufactured and sold $4.1 billion.

6. The Canadian Chemical Producers Association, *1991 Sector Update, Financial Reports* (Ottawa, 1992.)

Notes to Chapter 12 – A Matter of Policy

1. Parliament of Canada, *Hansard 1979* (Ottawa: Queen's Printer), p. 2129-2130.
2. *The Calgary Herald*, December 8, 1979.

Notes to Chapter 13 - The Industry and the Environment

1. Canadian Bankers Association, *Sustainable Capital: The Effect of Environmental Liability in Canada on Borrowers, Lenders and Investors* (Toronto: 1991).

Notes to Chapter 14 - The Impact of the Petroleum Industry on Canada

1. John J. Barr, "The Impact of Oil on Alberta: Retrospect and Prospect," A.W. Rasporich, ed., *The Making of the Modern West: Western Canada Since 1945* (Calgary: University of Calgary Press, 1984), p. 97-103.
2. Employment and Immigration Canada, *Human Resources in the Upstream Oil and Gas Industry: Changes, Challenges, Choices*, (Ottawa: 1992).

Bibliography

A Word on Sources:
The following bibliography contains all the secondary sources used in the preparation of this manuscript. It also contains a good collection of material directly related to the history of the petroleum industry in Canada, many of which are specific in nature and augment this book. For additional information about sources see Endnotes.

Anderson, Allan. *Roughnecks and Wildcatters: Hundreds of Firsthand Exciting Stories . . . of "The Oil Patch."* Toronto: MacMillan of Canada, Division of Gage Publishing Limited, 1981.

Anderson, Kenneth E., Bill D. Berger. *Modern Petroleum.* Tulsa: Penn-Well Publishing Company, 1978.

Ball, Max W. *This Fascinating Oil Business.* New York: The Bobbs-Merrill Company, 1940.

Barr, John J. "The Impact of Oil on Alberta: Retrospect and Prospect," in A.W. Rasporich, ed., *The Making of the Modern West: Western Canada Since 1945.* Calgary: University of Calgary Press, 1984.

Barry, P.S. *The Canol Project: Adventure of the U.S. War Department in Canada's Northwest,* Edmonton: P.S. Barry, 1985.

Berger, Thomas R. *Northern Frontier, Northern Homeland: The Report of the Mackenzie Valley Pipeline Inquiry.* Ottawa: Minister of Supply and Services, 1977, Vol. 1.

Bott, Bob. *Mileposts: The Story of the World's Longest Petroleum Pipeline.* Edmonton: Interprovincial Pipe Line Company, 1989.

Bragha, François. *Bob Blair's Pipeline: The Business and Politics of Northern Energy Development Projects: New Edition with the Story Behind the Alberta Prebuild.* Toronto: James Lorimer & Company, Publishers, 1979.

Brantley, J.E. *History of Oil Well Drilling.* Houston: Gulf Publishing Co., 1971.

Breen, David. *Alberta's Petroleum Industry and the Conservation Board.* Calgary: University of Alberta Press, 1993.

British Petroleum Company Limited. *Our Industry, Petroleum: A Handbook Dealing with . . . The British Petroleum Company Limited.* London: The British Petroleum Company Limited, 1977.

Broadfoot, Barry and Mark Nichols. *Memories: The Story of Imperial Oil's First Century as Told by its Employees and Annuitants.* Toronto: Imperial Oil Limited, 1980.

Canada, National Energy Board. *Canadian Energy Supply and Demand, 1990-2010.* Ottawa: Minister of Supply and Services Canada, 1991.

Canada, Employment and Immigration Canada. *Human Resources in the Upstream Oil and Gas Industry: Changes, Challenges, Choices.* Ottawa: 1992.

Canadian Bankers Association. *Sustainable Capital: The Effect of Environmental Liability in Canada on Borrowers, Lenders and Investors.* Toronto: 1991.

Canadian Petroleum Association. *Public Consultation Guidelines for the Canadian Petroleum Industry.* Calgary: Canadian Petroleum Association, 1989.

Comfort, Darlene J. *The Abasand Fiasco.* Ft. McMurray: Darlene J. Comfort, 1980.

Davis, E.M. *Canada's Oil Industry.* Toronto: McGraw-Hill Company of Canada Limited, 1969.

de Mille, George. *Oil in Canada West: The Early Years.* Calgary: George de Mille, 1969.

Doern, G. Bruce and Glen Toner. *The Politics of Energy.* Toronto: Methuen.

Dott, Robert H., et al. *Sourcebook for Petroleum Geology.* Tulsa: The American Association of Petroleum Geologists, 1969.

Easterbrook, W.T. and Hugh G.J. Aitken. *Canadian Economic History.* Toronto: Gage Publishing Limited, 1980.

Environment Council of Alberta. *Oil and Gas in Alberta: An Uncertain Future, Energy and Non-Renewable Resources Sub-Committee, Public Advisory Committees to the Environment Council of Alberta.* Edmonton: Environment Council of Alberta, 1989.

Ferguson, Barry Glen. *Athabasca Oil Sands: Northern Resource Exploration 1875-1951.* Edmonton: Alberta Culture/Canadian Plains, Research Centre, 1985.

Finch, David. *Traces Through Time: The History of Geophysical Exploration for Petroleum in Canada.* Calgary: Canadian Society of Exploration Geophysicists, 1985.

—. *Dealmakers: Canadian Petroleum Landmen and their Association.* Calgary: Canadian Association of Petroleum Landmen, 1989.

Fitzgerald, J. Joseph. *Black Gold With Grit: The Alberta Oil Sands.* Sidney, British Columbia: Gray's Publishing Ltd., 1978.

Foran, Max. "Calgary, Calgarians and the Northern Movement of the Oil Frontier," *The Making of the Modern West: Western Canada Since 1945*, by A.W. Rasporich, ed. Calgary: University of Calgary Press, 1984.

Foster, Peter. *The Blue-Eyed Sheiks: The Canadian Oil Establishment.* Toronto: Collins, 1979.

—. *The Sorcerer's Apprentices: Canada's Super-Bureaucrats and the Energy Mess.* Toronto: Collins, 1982.

—. *Other People's Money.* Toronto: Collins, 1983.

Foster, Peter. *From Rigs to Riches: The Story of Bow Valley Industries Ltd.* Calgary: Bow Valley Industries Ltd., 1985.

Gould, Ed. *Oil: The History of Canada's Oil and Gas Industry.* Victoria: Hancock House Publishers, 1976.

Gray, Earle. *The Great Canadian Oil Patch.* Toronto: Maclean-Hunter Limited, 1970.

Gray, Earle. *Wildcatters: The Story of Pacific Petroleums and Westcoast Transmission.* Toronto: McClelland and Stewart Limited, 1982.

Hanson, Eric J. *Dynamic Decade.* Toronto: McClelland and Stewart Limited, 1958.

Hilbourn, James D., consulting editor. *Dusters and Gushers: The Canadian Oil and Gas Industry.* Toronto: Pitt Publishing Company Limited, 1968.

House, J.D. *The Challenge of Oil: Newfoundland's Quest for Controlled Development.* Social and Economic Studies Number 30, Institute of Social and Economic Research. St. John's, Newfoundland: Memorial University, 1985.

Hurd, Vincent N. "Petrochemicals," in James D. Hilborn, ed., *Dusters and Gushers: The Canadian Oil and Gas Industry.* Toronto: Pitt Publishing Company, 1968.

Imperial Oil Limited. *Third Submission to the Restrictive Trade Practices Commission on the State of Competition in the Canadian Petroleum Industry.* Toronto, 1983.

Ingram, Robert L. *The Bechtel Story: Seventy Years of Accomplishment in Engineering and Construction.* San Francisco: Bechtel, 1968.

Kennedy, Tom. *Quest: The Search for Arctic Oil.* Edmonton: Reidmore Books, 1988.

Kerr, Aubrey. *Atlantic 1948 No. 3.* Calgary: S.A. Kerr, 1986.

—. *Corridors of Time.* Calgary: S.A. Kerr, 1988.

—. *Leduc.* Calgary: S.A. Kerr, 1991.

Kilbourn, William. *Pipeline: Transcanada and the Great Debate: A History of Business and Politics.* Toronto, Vancouver: Clarke, Irwin and Company Limited, 1970.

Laxer, James. *The Energy Poker Game: The Politics of the Continental Resource Deal.* Toronto, Chicago: New Press, 1970.

Laxer, James and Anne Martin. *The Big Tough Expensive Job: Imperial Oil and the Canadian Economy.* Toronto: Press Porcépic, 1973.

Leacock, Stephen. "In Praise of Petroleum," *Imperial Oil Review,* August/September issue, 1930.

Lewington, Peter. *No Right-of-Way: How Democracy Came to the Oil Patch.* Markham, Ontario: Fitzhenry & Whiteside, 1991.

Livingston, John. *Arctic Oil: The Destruction of the North?* Toronto: Canadian Broadcasting Corporation, 1981.

Lyon, Jim. *Dome - The Rise and Fall of the House That Jack Built.* Toronto: Macmillan of Canada, 1983.

Lysyk, Kenneth M. and Edith E. Bohmer, Willard L. Phelps. *Alaska Highway Pipeline Inquiry.* Ottawa: Supply and Services Canada, 1977.

Masters, John A. *The Hunters: Searching for Oil and Gas in Western Canada.* Calgary: Canadian Hunter Exploration Ltd., 1980.

McGillivray, The Honourable Mister Justice A.A. and L.R. Lipsett Esq. *Alberta's Oil Industry: Report of the Royal Commission appointed . . . to Inquire into Matters Connected with Petroleum and Petroleum Products.* Edmonton: Queen's Printer, 1940.

McRory, Robert E. *Energy Heritage: Oil Sands and Heavy Oils of Alberta.* Edmonton: Alberta Energy and Natural Resources, 1982.

National Energy Board. *Reasons for Decision, Northern Pipelines.* 3 vols. Ottawa: Minister of Supply and Services, 1977.

Nickle's Daily Oil Bulletin: 50th Anniversary Edition. Calgary, Nickle's Daily Oil Bulletin: 1988.

Nordegg, Martin. *The Fuel Problem of Canada.* Toronto: The MacMillan Co. of Canada Limited, 1930.

Peacock, Donald. *People, Peregrines and Arctic Pipelines: The Critical Battle to Build Northern Canada's Gas Pipelines.* Vancouver: J.J. Douglas Ltd., 1977.

Pearse, Peter H., ed. *The Mackenzie Pipeline: Arctic Gas and Canadian Energy Policy.* Toronto: McClelland and Stewart, 1974.

Pratt, Larry. *The Tar Sands: Syncrude and the Politics of Oil.* Edmonton: Hurtig Publishers, 1976.

Proctor, R.M. et al. *Oil and Natural Gas Resources of Canada.* Ottawa: Geological Survey of Canada, Energy Mines and Resources Canada, 1983.

Purdy, G.A. *Petroleum: Prehistoric to Petrochemicals.* Vancouver, Toronto, Montreal: Copp Clark Publishing Company, 1957.

Rasporich, A.W. "Oil and Gas in Western Canada, 1900-1980," *Canada's Visual History, Volume 70,* National Museum of Man, Ottawa, 1970.

Government of British Columbia. *Report of the Commission of the Honourable Justice A.A. Macdonald Relating to the Petroleum and Coal Industries.* Victoria, British Columbia: Petroleum Products Commission, Government of British Columbia, Vol. 1, The Petroleum Industry, 1936.

Roberts, Wayne. *Cracking the Canadian Formula: The Making of the Energy and Chemical Workers Union.* Toronto: Between the Lines, 1990.

Ross, Victor. *Petroleum in Canada.* Toronto: Southam Press Ltd., 1917.

Schaffer, Ed. *Canada's Oil and the American Empire.* Edmonton: Hurtig Publishers, 1983.

Shell Oil Company of Canada Limited. *Canadian Petrochemical Industry.* Toronto: Ryerson Press, 1956.

Shell Oil Company of Canada, Chemical Division. *The Canadian Petrochemical Industry.* Toronto: The Ryerson Press, 1956.

Smith, Philip. *The Treasure-Seekers: The Men Who Built Home Oil.* Toronto: Macmillan, 1978.

Solomon, Lawrence. *Energy Shock: After the Oil Runs Out.* Toronto: Doubleday Canada Ltd., 1980.

Spitz, Peter H. *Petrochemicals: The Rise of an Industry.* New York, Chichester, Brisbane, Toronto, Singapore: John Wiley and Sons, 1988.

Stenson, Fred. *Waste to Wealth: A History of Gas Processing in Canada.* Calgary: Canadian Gas Processors Association/Canadian Gas Processors Supplier's Association, 1985.

Victor, Ross. *The Evolution of the Oil Industry.* New York: Doubleday, Page and Company, 1920.

Walker, Michael, ed. *Oil in the '70s: Essays on Energy Policy.* Vancouver: The Fraser Institute, 1977.

Watkins, Campbell and Michael Walker, eds. *Oil in the Seventies: Essays on Energy Policy.* Vancouver, British Columbia, The Fraser Institute, 1977.

Timeline

Note: Page numbers indicate location of information concerning the event.

1700s

1719	Cree Wa-Pa-Sun gave sample of oil sand to Henry Kelsey of HBC 71
1778	Peter Pond saw oil sands deposits 71
1789	Alexander Mackenzie noted oil seeps along the Mackenzie River 40

1800s

1807	Gas derived from coal lit lamps in London streets 25
1821	Fredonia, New York, utilized wooden natural gas pipeline 25
1836	Coal gas streetlighting made its Canadian debut in Montreal 25
1846	Dr. Abraham Gesner made an illuminating oil from coal 102
1851	Charles Nelson Tripp incorporated Canada's first oil company 28
1853	Early Canadian gas pipeline built in Quebec 49
1854	International Mining and Manufacturing Company incorporated 28
1855	Canadian oil displayed at Paris Universal Exhibition 29
1857	J.M. Williams and Company incorporated 29
1858	James Miller Williams dug first oil well in North America 29
1859	H.C. Tweedle found oil and gas in Dover field, New Brunswick 32
1860s	Steam-powered rigs arrived from the United States 30
1862	First Canadian gusher blew in on February 19, in Ontario 29
1862	Oil pipeline connected Petrolia oilfield to Sarnia refinery 49
1866	Sour gas greeted drillers near Port Colborne, Ontario 32
1866	Pierre Berthelot chemically derived hydrocarbons 20
1867	Dominion of Canada introduced an import tariff on oil 103
1868	Canada first exported refined products 103
1877	Dmitri Mendeleef suggested percolating water for hydrocarbons 20
1880s	Canada became a net importer of oil 30
1880	16 Canadian refiners joined to form Imperial Oil Company 104
1883	Gas discovered at CPR Siding No. 8, near Medicine Hat, Alberta 32
1889	First producing gas well in Essex County, Ontario 32
1889-90	Discoveries of natural gas in Ontario 129
1890	Well drilled near Medicine Hat hit gas 32
1891	Canada first exported natural gas to Buffalo, N.Y. 32
1895	Natural gas began flowing from Ontario to the United States 49
1897	Pelican Rapids struck gas and blew wild 33

1898	Imperial Oil became part of Standard Oil 106
1898	First automobile arrived in Canada 30, 110

1900s

1901	Rocky Mountain Drilling Company organized 33
1904	Discovery of the Medicine Hat gas field 33
1904	Canadian Oil Company Limited formed 104
1906	British American Oil Company formed 104
1908	Imperial Oil opened the first Canadian gas station 110
1909	Old Glory struck gas well near Bow Island, Alberta 33
1909	New Brunswick's first gas well came in at Stoney Creek 32

1910s

1910	Calgary Natural Gas Company formed 33
1911	Union Gas Company of Canada Limited formed 32
1911	Jim Cornwall noticed oil on the Mackenzie River 40
1912	Canadian Western Natural Gas Company built 280-km gas line 33
1912	Eugene Coste built the first major gas pipeline 49
1914	Calgary Petroleum Products #1 well hit oil in Turner Valley 36
1914	Oil fever swept Calgary 36
1914	First Canadian oil absorption plant built at Turner Valley 62
1914	First true gas processing plant built in Canada at Turner Valley 39, 62
1917	Ontario natural gas production peaked 48
1917	First battery of condensation stills built at Montreal 106
1919	British Columbia withdrew its lands from exploration 48
1919	Imperial Oil began exploratory drilling on the Mackenzie River 42

1920s

1920	Turner Valley refinery burned down 62
1920	World's most northerly well struck oil at Norman Wells 42
1921	Canadian Western bought processed Turner Valley gas 37
1923	Imperial facility in Calgary began refining Turner Valley oil 105
1923	First continuous thermal cracking stills built at Calgary 106
1923	Viking-Kinsella gas field justified a pipeline to Edmonton 44
1924	Royalite Oil built a gas "sweetening" plant in Turner Valley 146
1924	First gas sweetening plant in Canada built at Port Alma, Ontario 62
1924	Gas caught fire at Royalite #4 well, October 1937
1924	Union Gas Co. built a processing plant "sweeten" gas 146
1925	Rotary drilling introduced to Turner Valley 39
1925	International Bitumen produced pure bitumen at Bitumount 73

1925	Discovery of heavy oil at Wainwright 44
1926	Introduction of tetra-ethyl lead in gasoline 106
1927	McLeod #2 well introduced nitro-shooting to Canada 39
1927	Imperial owned 90 percent of Canada's refining capacity 106
1927	McColl Brothers and Frontenac merged into McColl-Frontenac 106

1930s

1930	Max Ball and B.O. Jones formed Abasand 73
1932	Alberta government attempted to control gas waste 38
1932	Turner Valley Gas Conservation Board established 132
1933	Federal government imposed 3.7 cent/gallon protective duty 83
1934	Gas found at Lloydminster 44
1936	Turner Valley Royalties #1 struck oil, June 16 38
1936	"Royalties" system of financing wells introduced 40
1936	Refinery opened at Norman Wells, NWT 42
1938	Alberta Petroleum and Natural Gas Conservation Board began operating 39, 64
1938	Parliament passed a Resource Transfer Act 132
1939	Heavy oil discovered at Lloydminster 44

1940s

1940	Early catalytic cracker installed at Sarnia, Ontario 106
1941	Tar sands mining began at Abasand plant 73
1942	Turner Valley field produced 10 million barrels of oil in one year 39
1944	Wet sour gas discovered at Jumping Pound, Alberta 44
1944	Gas storage began in Turner Valley 39
1944	Canol pipeline completed, abandoned in 1946 49
1946	Canol pipeline completed in 1944, abandoned in 1946 49
1946	Canada produced less than 10 percent of oil Canadians used 107
1947	February 13, Imperial Leduc No. 1 well struck oil 46, 133
1948	Redwater, Alberta field discovered 47
1948	Pincher Creek, Alberta gas plant began operating 47
1948	Atlantic Leduc #3 well blew wild for most of the year 39, 46-7
1948	Water injection introduced to Turner Valley oilfield 39
1949	Golden Spike, Alberta oilfield discovered 47
1949	Dinning Commission report released 51
1949	Ottawa passed Oil or Gas Pipe Lines Act 51, 134

1950s

1950	Oil found at Cessford, Alberta 47

1950	First leg of Interprovincial Pipeline completed 50
1951	Important light oil discovery at Virden, Manitoba 47
1952	First Canadian sulphur plants, Jumping Pound, Turner Valley, Alberta 64, 146
1952	Oil discovery at Westerose, Alberta 47
1952	Westcoast removed first gas from Alberta 53
1952	Alberta oil: Bindloss, Hussar, Minnehik, Duck Lake, Nevis, Olds 47
1952	Interprovincial Pipeline provided Alberta crude to Ontario 50, 120
1952	Canadian-Montana Pipeline built 134
1953	Oil moved in Trans-Mountain pipeline from Edmonton to Vancouver 50
1953	Pembina, Alberta oilfield discovered 47
1954	AGTL incorporated, issued public shares in 1957 55
1955	Boundary Lake, B.C. oil discovery 48
1956-57	Suez Crisis 134
1957	TCPL experienced a 5.5 km blowout of gas along its line 55
1957	Mackenzie delta was a focus of ground and air surveys 90
1957	Westcoast Transmission Limited gas pipeline completed 54, 65
1957	Alberta government began measuring gas plant emissions 147
1957	Arctic Circle exploration began in earnest 88
1957	Swan Hills, Alberta oilfield discovered 47
1957	First well drilled in the Yukon 88
1957	Gas began flowing through the TransCanada pipeline 55, 65
1958	Borden Commission report published 134
1958	TCPL completed as longest gas pipeline in world 55
1959	Cities Service Athabasca constructed oil sands refinery 74
1959	Ontario oil production exceeded record set in 1895 48
1959	National Energy Board formed 134

1960s

1960	Alberta set up an Air Pollution Control Branch 147
1961	National Oil Policy enacted by Diefenbaker government 135
1961-62	Winter, first well in the Arctic Islands 89
1961	Alberta gas began flowing to California 56, 134
1962	Rachel Carson published *Silent Spring* 145
1962	Drilling began in the Mackenzie Delta-Tuktoyuktuk Peninsula 90
1966	First well test the Beaufort Sea 91
1967	Arco Humble #1 wildcat spudded at Prudhoe Bay, Alaska 90
1967	Ontario oil production peaked 48
1967	Shell began drilling wells off West coast 98

1967	Suncor opened tar sands plant at Fort McMurray, Alberta 75
1968	Club of Rome formed 145
1968	Panarctic Oils formed 89
1968	Prudhoe Bay oil strike in Alaska 57
1969	Drake Point gas discovery, Arctic Islands 90

1970s

1970	Tanker *Arrow* grounded off Nova Scotia 145
1970	Quebec created a provincially-owned petroleum company 136
1971	Alberta government created Department of Environment 148
1971	First drilling on Labrador Shelf 96
1971	Alberta government enacted a Clean Air Act 148
1971	Ottawa established national air-quality objectives 148
1972	*Limits to Growth* published by The Club of Rome 145
1973	Foreign Investment Review Agency (FIRA) formed 138, 141
1974	Mackenzie Valley Pipeline Inquiry appointed 58
1974	Alberta Oil Sands Technology and Research Authority formed 81
1976	Interprovincial extended its pipeline to Montreal 120
1976	IPL reached its present length of 3 680 kilometres 50
1977	Arctic Marine Oilspill Program established 149
1977	Berger Commission urged 10-year Northern postponement 58, 148
1978	Syncrude began refining oil from tar sand 76
1979	Hibernia oil strike 97

1980s

1980	AGTL changed its name to NOVA Corporation 56
1980	National Energy Program announced in the budget October 28 139
1980	Canadian consumption finally began to decline 108
1982	Prebuild sections of Alaska Gas Pipeline opened 59
1985	Deregulation of industry began 50, 121
1985	Shell built first exclusively oil sand refinery 107
1985	Panarctic shipped 100 000 bbl of Bent Horn oil to Montreal 93
1985	Norman Wells pipeline opened 51
1985	Western Accord signed by Alberta, B.C. and Saskatchewan 141
1986	Cold Lake project began developing tar sand in situ 80
1986	Offshore development crashed as oil prices dropped 96
1986	Oil prices crashed 108, 161
1987	Canadian oil companies moved into Russian fields 100
1989	Exxon *Valdez* tanker ran aground in March 152

1990s

1991	534 km gas pipeline moved gas to Vancouver Island	56
1992	Balmoral field went into production, East coast offshore	96
1992	Anniversary of the Geological Survey of Canada	161
1992	Iroquois Gas Transmission System began moving gas to US	57
1992	Bi-Provincial Upgrader processed heavy oil at Lloydminster	85
1992	Canadian Association of Petroleum Producers formed	157

Index

A

Abasand Oils Limited 73
Absher, Jacob Owen 78
Acidizing 39
Adgo F-28 91
Air Pollution Control Branch, Alberta government 147
Alaska Highway 58, 148
Alaska Highway Pipeline 58
Alberta & Southern Gas Co. 56, 134
Alberta 14, 23, 105, 130, 135, 138, 141, 148
Alberta Department of Health 147
Alberta Energy Co. 138
Alberta Gas Ethylene 122
Alberta Gas Trunk Line (see also Nova) 55
Alberta Nitrogen Products 117
Alberta Oil Sands Technology and Research Authority, AOSTRA 81
Alberta Petroleum Industry-Government Environmental Committee 149
Alberta Petroleum Marketing Commission 138
Alberta Public Utilities Board 132
Alberta Research Council 72, 73
Alberta-to-California pipeline system 57
Allen's Process 103
Allied War Supplies Corporation 117
Alsands 77, 80
Altamont 57
Alyeska 57
Amauligak 93
American Navy 82
Amoco Canada Petroleum Co. 153, 163
Aquitaine 90, 136
Arco Humble #1 90
Arctic Gas pipeline 91
Arctic Islands 16, 17, 23, 89, 93
Arctic Marine Oilspill Program 149
Arctic Pilot Project 93
Arctic pipelines 18
Arctic Waters Pollution Prevention Act 148
Arrow tanker accident offshore of Nova Scotia 149
Ashland Oil 136
Athabasca, Alberta 71, 79
Athabasca oil sands, Alberta 69, 107
Athabasca River, Alberta 71, 73, 88
Atlantic Leduc #3 39, 47
Atlantic Richfield 76
Avalon and Jeanne d'Arc basins 97

B

B.C. Petroleum Corp. 137
BA-Shawinigan 120
Ball, Max 73
Balmoral field, East coast offshore 96
Banff, Alberta 64, 145
Barber Asphalt and Paving Company 78
Barr, John 158
Bathurst island, Arctic Islands 89
Beart, Robert 31
Beaufort-Mackenzie Delta 23
Beaufort Sea 16, 23, 24, 90-93, 145, 149, 160
Beaumont, Texas 31
Bell, Robert 71
Bennett Buggy 13
Bent Horn oil field, Arctic Islands 90, 93
Berger Commission 58
Berger, Thomas 58, 148
Berthelot, Pierre 20
Bertie-Humberstone field in Welland, Ontario 32
Bertrand Allegations 113
Bertrand, Robert 112
Bi-Provincial Upgrader 84, 85
Bindloss, Alberta 47
Bishop, Texas 119
Bitumen 69, 71
Bituminous Sand Extraction Company 78
Bitumount, Alberta 73
Black Creek, Ontario 28, 30
Black, Patricia, Alberta's energy minister 163-4
Blair, S.M. 72, 74
Blow-out preventers 95
Board of Transportation Commissioners 134
Bonnie Glen, Alberta 47, 88, 160
Borden, Manitoba 135
Borden, Robert 134
Bosworth Creek, NWT 42
Bosworth, T.O. 42
Boundary Lake, B.C. 48
Bow Island gas field, Alberta 37, 131
BP Canada 79, 112
Brander-Smith, David 153
British American Oil Company (see also Gulf Canada) 63, 106, 111, 120, 147
British North American Act 130
Broadbent, Ed 113, 129
Brown, John George (Kootenai) 33
Butane 64, 66

C

Cable tool drilling 28
Cabot Carbon of Canada Limited 118
Calgary 64, 118, 131, 160
Calgary Brewing and Malting Company 33
Calgary Herald viii
Calgary Petroleum Products Company 36
Cameron Brook, Alberta 33
Cameron Island, Arctic Islands 90
Canada Development Corporation 135-6, 141
Canadian Arctic 77
Canadian Arctic Gas 93, 148
Canadian Association of Petroleum Producers 157
Canadian Bankers Association 143
Canadian Chemicals Company Limited 119
Canadian Delhi Oil Company 54
Canadian drilling rig 30
Canadian Fracmaster Ltd. 100, 156
Canadian Industries Limited 119
Canadian Marine Drilling Ltd. 91
Canadian-Montana Pipeline 134
Canadian Occidental Petroleum Co. 153
Canadian Offshore Oil Spill Research Association 149
Canadian Oil Company Limited 104, 106, 111
Canadian Pacific Railway 130
Canadian Petroleum Association vii-ix, 149, 157
Canadian Petroleum Products Institute 153
Canadian Western Natural Gas 36, 63, 131
CanaKaz Global Oils Inc. 156
Canol pipeline 15, 43, 106
Canterra Energy 90, 136
Cape Allison, Arctic Islands 90
Caracas, Venezuela 136
Carbon-black plant 131
Carney, Pat 141
Caroline, Alberta 68, 151, 153
Carson, Rachel 145
Carswell Gastrobleme 22
Casing 28
Caspian Sea 156
Catalytic cracking 105
Celanese Corporation of America 119
Cessford, Alberta 47
Chevron 97
China 134
Chinese 25, 28

Chipewyan Indians 71
Chretien, Jean 148-9
Christie, William 88
Cisco, Arctic Islands 90
Cities Service Athabasca Inc. 74, 76
Clark, Dr. Karl 72, 74, 78
Clark, Joe 16, 127-9, 135, 139-141
Clean Air Act, British version 144
Clearwater River, Alberta 71
Cleveland, Rod 155
Clover Bar, Alberta 119
Club of Rome, The 136, 141, 145, 151
Coal bed gas 56
Cochin pipeline 60
Cochrane to Edmonton (Co-Ed) pipeline 60
Cogeneration plants 67
Cohasset, East coast offshore 15, 96
Cohasset A-52 96
Cohasset well, D-42 95
Cold Lake, Alberta 79, 80, 85
Combines Investigation Bureau 112
Commonwealth Air Training project, WWII 36, 107
Commonwealth of Independent States 156
Competition in the Canadian Petroleum Industry 113
Conservation, economic 144
Consumers Association of Canada 112
Consumers' Cooperative Refinery 83
Cordillera 23
Cornwallis island, Arctic Islands 89
Coste, Eugene 20, 32, 49
Cracking 104, 105
Cree Indian Wa-Pa-Sun 71
Cretaceous formation 95
Cretaceous sandstone 36
Crosbie, John 129

D

da Vinci, Leonardo 31
Dawson Creek 160
Dawson, Dr. George M. 32-4
Dealmakers viii
Dempster Highway 148
Department of Indian and Northern Affairs 148
Department of the Interior 132
Depression 13, 14
Deregulation of petroleum industry in 1985 50, 67
Detroit River, Michigan-Ontario 32

Devonian Age 23
Diefenbaker, John 89, 134, 135
Dingman, Archibald Wayne 33, 36
Dingman, C.W. 72
Dinning Natural Gas Commission 51, 133, 134
Dirty Thirties 13
Dome Petroleum 89, 91, 121, 122
Dominion Tar and Chemical Company (today known as Domtar) 120
Domtar, Dominion Tar and Chemical Company 120
Donald, Ivan vii
Dover oil and field near Moncton, New Brunswick 32
Dow Chemical of Canada 118, 121, 122
Downstream - definition of the term 101
Drake, Edwin 15, 29
Drilling, "Jack up" rigs 94, 96
Drilling, cable tool drilling 28
Drilling, Canadian rig 30
Drilling, rotary 31, 39
Drilling, semi-submersible platforms 94
Drilling, technological advances in the art of 30
Drillships 94
Duck Lake, Alberta 47
Duguid, A.T. vii
Dynamic Decade, 159

E
Eagle Plains, Yukon 89
East Coast developments 16, 99, 140
East Coast Spill Response Inc. 151
Economic conservation 144
Edmonton, Alberta 122, 131, 132, 160
Edmonton Eskimos 159
Elder, William 36
Eldorado Gold Mines, NWT 42
Elk Point, Alberta 80
Ells, Dr. Sidney C. 72, 78
Empress, Alberta straddle plant 60
Energy Resources Conservation Board 77, 133, 144, 147, 149, 151, 153, 154
Energy self-sufficiency tax 128
Enniskillen Township, Ontario 28, 103
Environment Canada 149
Environmental audits 143
Environmental concerns 145
ERCB see Energy Resources Conservation Board
Essex County, Ontario 32

Estevan, Saskatchewan 160
Ethane 66
Ethel Lake, Alberta 80
Euphrates River 25
Europe 25
Exportation of Power and Fluids Act 129
Exshaw, Alberta 64
Exxon 98, 152
Exxon *Valdez* oil spill 98

F
Fairbanks, Alaska 43
Fenn-Big Valley, Alberta 47, 160
Fertilizer 119
Finch, David viii-x
FIRA - Foreign Investment Review Agency 138, 141
Fitzsimmons, R.C. 73
Foothills Pipelines Ltd. 93
Foreign Investment Review Agency - FIRA 138, 141
Forestry, Oil and Gas Review viii
Fort McMurray, Alberta 72-3, 160
Fort Norman, NWT 42
Fort Saskatchewan, Alberta 118, 122
Fort St. John, B.C. 48, 160
Fortier, Yves 89
Frasch, Herman 104
Fraser, A.W. 71
Fraser River, B.C. 50
Fredonia, New York 25
Free Trade Agreement, U.S.-Canada 125
Frontenac Oil Refineries 106
Frontier Drilling Costs, 1960 - 1990 99

G
Gas, "Sour" 146
Gas and Oil Products Ltd. 63, 106
Gas conservation plants 64
Gas, natural gas 17, 61
Gas Resources Preservation Act 134
Gas, waste of natural gas 131
Geological Survey of Canada 15, 32, 71, 88, 89, 156, 161
Geology, inorganic school of 20
Gesner, Dr. Abraham 14, 102
Ghadafy, Colonel Moammar 136
Gold, Thomas 21
Golden Spike, Alberta 47, 160
Gordon Royal Commission on Canada's Economic Prospects 134-36

Gordon, Walter 135
Grand Banks 24, 95-97, 140, 145
Grande Prairie, Alberta 160
Gravity Based System, Hibernia 97
Great Bear Lake, NWT 42
Great Canadian Oil Sands 75
Great Pipeline Debate 54
Great Slave Lake 42
Greece 25
Greenland 97
Gulf Canada Resources Ltd. (see also British American) 8, 77, 83, 92-3, 97-8, 100, 107, 111-2, 120, 156
Gulf War 12
Gulless, Micky vii

H
Halifax 106
Hanson, Eric 159
Hartell, Alberta 63, 106
Heavy oil 17, 69, 81
Hecate Strait, B.C. 98
Hell's Half-Acre, Turner Valley, Alberta 38
Henry, J.T. 102, 103
Heritage Savings Trust Fund, Alberta 161
Herron, Bill 36
Hibernia, East coast offshore 15, 24, 97, 98, 140
Hillsborough #1, P.E.I. 94
Home Oil 38, 135, 136
Houdry, Eugene 105
House of Commons, Ottawa 127
Howe, Bruce 156
Howe, C.D. 40, 54
Hudson Bay 90
Hudson Bay Basin 23
Hudson's Bay Company 71, 88
Hunt, Thomas Sterry 15
Hunt, Warren 21
Hurricane Hydrocarbons Inc. 156
Husky Oil Ltd. 68, 82-84, 90, 151
Hussar, Alberta 47
Hutchins, Bob vii
Hydrogen sulphide 62, 146

I
Ice islands 90
Iceberg Alley 97
Icebergs 97
Immerk 13-48 91

Imperial Oil 45, 79, 80-3, 104-6, 107, 110-12, 117-18, 160, 163
Independent Petroleum Association of Canada 157
Industrial Revolution 27
Institute of Sedimentary and Petroleum Geology 156
International Bitumen Company 73
Interprovincial Pipe Line 50, 54, 60, 82, 120
Investment Canada 141
Iran 139
Irving Oil 107, 110-11
Island Development Company 94
Iso-butylene 105

J
Jack-up rig 94, 96
Jaremko, Gordon viii, x
Jasper Park, Alberta 72
Joffre, Alberta 122
Jones, B.O. 73
Jumping Pound, Alberta 44, 64, 146

K
Kazakhstan 156
Kelsey, Henry 71
Kerosene 102
King Idris 136
Kitimat, B.C. 122
Koch Oil Co. 163
Korean War 50
Kulluk, drilling vessel 92

L
Labrador 94, 96
Labrador Coast 96
Labrador Shelf 95
Lake Ainslie, Nova Scotia 32
Lake Erie 28
Lake St. Clair 29
Lambton County, Ontario 29
LASMO plc. 96
Lawrence, Allan 113
Leacock, Stephen 12
Leduc #2 46
Leduc #1 45-6
Leduc, Alberta 45, 133, 134, 146, 158-160
Leduc gas conservation plant 60
Leduc-Woodbend field, Alberta 119
Leitch, Merv 128
Lemming, Alberta 80
Les pères de la Fraternité Sacerdotale 48

Lethbridge, Alberta 131
Libya 136
Limits to Growth, by Club of Rome 151
Lindbergh, Alberta 80
Lineham, John 33
Link, Ted 42
Little Cornwallis Island, Arctic Islands 93
Lloydminster, Alberta-Saskatchewan border 44, 82, 85, 160
Lodgepole, Alberta 153
Logan Canyon sands 95
Longview, Alberta 63
Lougheed, Peter 128, 129, 136, 138, 140, 159
Lumley Ed, 127-9

M

MacDonald Hotel, Edmonton 158
Macdonald, John A. 130
MacGarvey, W.H. 15, 30
Maciej, Hans vii, ix
Mackenzie, Alexander 40, 71
Mackenzie Delta, NWT 58, 90, 93
Mackenzie River, NWT 40, 87
Mackenzie Valley Pipeline Commission 148
Mackenzie Valley Pipeline Research Group in 1969 57
MacLeod, Jack 151-2
Macoun, John 71
Madison Natural Gas plant, Turner Valley 64
Maine 120
Manitoba 130, 159, 160
Manning, Premier 158
Maple Leaf line 93
Maritimes 13
May Lake, Alberta 80
McColl Brothers Limited 106
McColl-Frontenac 106, 111
McConnell, R.G. 40, 71
McKenzie-Brown, Peter viii-x
McKinnon, Ian 134
McLellan, William R. S. x
McMahon, Frank 53, 98
Medicine Hat, Alberta 122, 131
Melville Island, Arctic Islands 89
Mendeleef, Dmitri 20
Mildred Lake, Alberta 74, 76
Minnehik, Alberta 47
Mitchell, Gordon 156
Mitsue, Alberta 89
Mobil Oil Canada 94-5, 97

Molikpaq, Gulf Canada's 90-metre square floatable island 92
Montreal 25, 106, 119, 120, 123, 135
Morinville, Alberta 119
Mulroney, Brian 141, 153
Murphy Oil Corp. 98

N

N-butylene 105
Nassichuk, Walter 156
National Energy Board 56, 58-9, 134-6 141-4, 148, 161
National Energy Program 16, 113, 127-8, 139-41, 155, 157, 160
National Energy Program 80
National Oil Policy 134, 135
National Policy, John A. Macdonald's 130
Natural Gas Exporter viii
Natural Gas Intelligence viii
NEB - see National Energy Board
NEP - see National Energy Program
Nevis, Alberta 47, 147
New Brunswick 15, 32, 105
New Brunswick albertite 102
New York Mercantile Exchange 157
Newfoundland 15, 24, 95-97, 140, 145
Newfoundland's Grand Banks 15
Nielson, Glenn 82
Nipisi, Alberta 89
Noah's ark building instructions from Genesis 25
Norcen Energy Resources Ltd. 156
Norman Wells, NWT 15-16, 40, 43, 90, 93, 106-107
North American Kerosene Gas and Lighting Company 102
North Korea 134
North Sea 140, 157
Northwest Territories 15, 23, 99
Northwestern Utilities Limited 64
NOVA Corporation of Alberta - see also Alberta Gas Trunk Line 55-7, 121-3, 125, 163
Nova Scotia 15, 95, 140, 145, 149
Nova Scotia Resources Limited 96
Novacor Chemicals 122
Nuclear explosion, plans to liquify tar sands with 79

O

Octane rating 105
Offshore wells 94
Ohio 30

Oil and Gas Conservation Board, Alberta 76
Oil and Gas Production and Conservation Act, Ottawa 148
Oil City, Alberta 33
Oil Creek, Ontario 29
Oil or Gas Pipe Lines Act, 1949, Ottawa 51, 134
Oil sands 14, 17, 69, 73, 138
Oil sands plants 73
Oil spill, Exxon *Valdez* 98
Oil spills 145
Oil Springs, Ontario 9, 14, 29, 103
Oilweek magazine vii
Old Glory gas well near Bow Island, Alberta 33
Old Nig, ox named 42
Olds, Alberta 47
Onandaga E-84 95
Ontario 15, 23, 105, 131-3
OPEC - see Organization of Petroleum Exporting Countries
Operation Franklin 89
Oracle of Delphi 25
Organization of Petroleum Exporting Countries 14, 76, 108, 121, 136-7, 141, 157, 162
OSLO project (for Other Six Lease Operations) 77
Ottawa 127
Ottawa Valley Line 135
Ottawa's Department of Energy 163

P
Pacific Gas and Electric Co, of San Francisco 56, 134
Pacific Gas Transmission Co. Shipments 134
Pacific Petroleums 112
Paleozoic 23
Panarctic Oils 89-91, 93, 135
PanCanadian Petroleum Ltd. vii, 156
Pangaea 22
Panuke B-90 96
Paris Universal Exhibition, 1855, Canadian oil displayed at 29
Peace River, Alberta and B.C. 48, 79, 80
Pearson, Lester 135
Peel Plateau Exploration 88
Peking 25
Pelican Portage, Athabasca River, Alberta 71
Pembina, Alberta 47
Pennsylvania 15
Pentanes plus 64

Petro-Canada 83, 96-7, 110-11, 112, 129, 135, 138, 163
Petrochemicals 115
Petrofina 112
Petroleum and Natural Gas Conservation Board, Alberta 64, 133
Petroleum, basis for the word 25
Petroleum History Society, The vii-ix
Petroleum Incentives Program 99, 129, 139-40
Petroleum Industry Training Service 149
Petroleum Monitoring Agency 113
Petroleum, organic theories of petroleum origin 20
Petroleum, origin of 20
Petroleum, origin of in Canada 22
Petroleum Services Association of Canada 157
Petroleum, use of 24
Petrolia, Ontario 30, 117, 130
Petrosar Ltd. 122
Pew, John Howard 75
Pincher Creek, Alberta 33, 47, 147
Pipeline technology 16
Plains Upgrader 83
Poland 15
Polar Gas Pipeline Project 59, 93
Politics of Natural Gas Transportation 51
Polymer (later called Polysar) 117, 118, 123, 125
Polymerization 105
Polysar (first called Polymer) 117, 118, 123, 125
Pond, Peter 71
Port Colborne, Ontario 32
Port Mann, B.C. 50
Prairie Regional Oil Spill Containment and Recovery Advisory Committee 149
Precambrian Shield 23
Priddle Principle 141
Priddle, Roland 141
Prince Edward Island 15
Prince Edward Island's Hillsborough Bay 15
Prince Rupert, British Columbia 119
Privy Council of Britain's House of Lords 130
Project on the Predicament of Mankind 145
Propane 64, 66
PROSCARAC 149
Prudhoe Bay 57, 76, 77, 90, 93
Public Review Panel on Tanker Safety and Marine Spills Response 153

Q
Quebec 15
Queen Charlotte Islands, B.C. 98
R
Rainbow, Alberta 89
Rankin, Alex vii
Reagan, Ronald 141
Red Deer, Alberta 160
Redwater, Alberta 47, 77, 160
Refinery openings and closures in Canada 109
Regina, Saskatchewan 106, 131
Restrictive Trade Practices Commission 113
Ribstone, Alberta 82
Richards Island, Arctic Islands 91
Richfield Oil Company 79
Robinson, Terry vii
Rodrigues, Stephen vii
Rome 25
Ross, Victor 12
Rotary drilling 31, 39
Rowland, Les vii
Royal, Darrel 159
Royal Dutch Shell 141
Royalite #4 well 37, 146
Royalite Oil Company 36, 62-3, 117, 131, 146
Ruffner brothers, David and Joseph 28
Russia 14
S
Sable Island 24, 59, 94, 95
Sable Island C-67 well 95
Safety Boss 12
Sarnia, Ontario 30, 106-7, 117-120, 122-3, 130
Saskatchewan 22-3, 130, 133, 137-8, 141, 148, 159-60
SaskOil 83, 137
Saudi Arabia 157
Scotford, Alberta 107, 123
Scotian Shelf 24, 95, 96
Semi-submersible drilling platforms 94
Seneca Oil Company 29
Shaw, Hugh Nixon 29
Shawinigan Chemicals 120
Shell Canada vii, 44, 64, 68, 77-80, 83, 95-8, 106-7, 110-11, 118-19, 122-5, 146, 151
Shu Han Dynasty 25
Siljan Ring 21
Silliman, Dr. Benjamin 102, 104
Skagway, Alaska 15, 43
Skate, Arctic Islands 90
Slipper, S.E. 72
SOAP, Sarnia Olefins and Aromatics Project 122
Social Credit government of Alberta 133, 136, 148
SOQUIP 136
South Korea 134
Spencer, William 103
Spindletop well, Beaumont, Texas 31
Spokane, Washington 158
St. Clair River, Ontario 117
St. John's, Newfoundland 97
St. Laurent, Louis 16
Standard Oil Company 104-106
Stenson, Fred vii
Steveston Land and Oil Company, B.C. 98
Stoner, O.G. 113
Stoney Creek, New Brunswick 32
Stoney Indians 33
Subsidy of oil 13
Suez Crisis 134
Sulphur 64, 66, 68, 75, 103-5, 116, 133, 146, 147-8
Sun Oil Company 75, 110
Suncor 75, 160
Supertest 111
Supreme Court of Canada 137, 140
Sustainable development 68, 151
Swan Hills, Alberta 47, 88
Syncrude Canada Ltd. 17, 74, 76-78, 85, 138, 158, 160
Synthetic oil production and technology 17
T
Tanner, Nathan 46, 74
Tarsuit Island, Beaufort Sea 92
Teheran 136
Terra Nova, East coast offshore 97
Texaco Canada 106, 107, 111, 112
Texas Railroad Commission 133
Thermal cracking 104
Thomson, Jerry 157
Tillbury, Ontario 131
Tippett, Clint vii
Titusville, Pennsylvania 29
Toledo, Ohio 32
Toronto 15, 131
Toronto Stock Exchange 128, 163
Traces Through Time viii
TransCanada PipeLines Limited 16, 54, 65, 134

Transportation 51
Trinidad bitumen 102
Tripp, Charles Nelson 28
Trois-Riverès, Quebec 15, 48
Trudeau, Pierre, 16, 129, 137-140
Turner, John 140
Turner Valley 31, 35, 39, 62, 106, 117, 131-2, 146
Turner Valley Firsts 39
Turner Valley Gas Conservation Board 132
Turta, Mike 45
Tweedle, Dr. H.C. 32

U
Ukraine 15
Ultramar 110
Underground Test Facility 81
Uniacke G-72 95
Union Carbide Canada 120, 122
Union Gas Company of Toronto 32, 62, 146
United Farmers of Alberta 132
United States Atomic Energy Commission 79
United States Federal Power Commission 53, 58
Upgraders 83
Upstream - definition of term 101
Ur 25
Uzen 156

V
Valdez, Alaska 57, 152, 153
Valdez, Exxon tanker ran aground in March of 1989 152
Valdez pipeline, Alaska 93
Vancouver, B.C. 50, 106
Viking-Kinsella gas field, Alberta 44, 131
Volck, Johan 25
von Hamerstein, Alfred 72

W
Wa-Pa-Sun, Cree Indian 71
Wabasca, Alberta 79
Wainwright, Alberta 44, 82
Wainwright Producers and Refiners Ltd 106
Warplanes 105
Wartime Oils Ltd. 40, 130
Waste of crude oil 38
Waterways, Alberta 73
Watson Lake, Yukon 43
Wega-D Geophysical Ltd. 156
West Coast 98
West Venture 95

West Venture N-91 95
West Virginia 28
Westcoast Transmission Limited natural gas pipeline 48, 53, 65
Western Accord 141, 155, 157, 158
Western Canada Basin - see Western Canada Sedimentary Basin
Western Canada Sedimentary Basin 16, 18, 23, 88, 160
Western Minerals 88
Western Oil and Coal Company 34
Western Pipelines 54
Western Sedimentary Basin - see Western Canada Sedimentary Basin
Westerose, Alberta 47
Weyburn, Saskatchewan 160
Whale oil 14, 25, 27, 102
Whitehorse, Yukon 15, 43, 89, 106, 107
Whiterose, East coast offshore 97
Wignall, Christine vii
Williams, James Miller 14, 29, 103
Winchell, Dr. A. 29
Winnipeg, Manitoba 131
Winter Harbour #1, Arctic Islands 89
Wizard Lake, Alberta 47, 88, 160
Wolf Lake, Alberta 80
World War II 12-15, 36, 111

Y
Yom Kippur War 137
Young, James 102
Yukon 15, 23, 88, 99

This book is to be returned on or before
the last